THE PHYSICAL BEHAVIOUR OF MACROMOLECULES WITH BIOLOGICAL FUNCTIONS

MONOGRAPHS IN
[*MOLECULAR BIOPHYSICS AND BIOCHEMISTRY*]

Edited by
H. Gutfreund,
Department of Biochemistry,
University of Bristol

The Physical Behaviour of Macromolecules with biological functions
S. P. Spragg

THE PHYSICAL BEHAVIOUR OF MACROMOLECULES WITH BIOLOGICAL FUNCTIONS

S. P. Spragg
Department of Chemistry,
University of Birmingham

JOHN WILEY & SONS

Chichester · New York · Brisbane · Toronto

British Library Cataloguing in Publication Data:

Spragg, S. P.
The physical behaviour of macromolecules with biological functions. —(Monographs in molecular biophysics and biochemistry).
1. Macromolecules—Measurement
I. Title II. Series
574.1′924 QP801.P64 80-40280

ISBN 0 471 27784 3

Photosetting by Thomson Press (India) Limited, New Delhi
Printed and bound in England by The Pitman Press Limited, Bath, Avon

To
Daphne, Carolyn, and Martin

for tolerating my periods of withdrawal

. . . anything done unwisely cannot be considered a skill . . .

From King Alfred's epitaph (A.D. 849–899).

Series preface

The term 'molecular biophysics' means different things to different people. To some the subject matter is covered by X-ray structure studies of molecules of biological interest, while others believe that the application of one of several different physical techniques to the study of biological systems would come under that title. Personally, I consider Karl Pearson's definition most appropriate: 'Biophysics is the study of the physical laws which govern the behaviour of biological systems'. Obviously, physical techniques are usually required to study physical laws, but the emphasis—if one considers biophysics as part of biological science—must be on the 'behaviour of a biological system'. All too often techniques have ended up as 'methods in search of a problem'. One could remark, somewhat cynically, that while most techniques have a limited life, fundamental biological problems never do get solved. This is probably true because people ask open-ended questions like 'what is the mechanism of enzyme action?' or 'how does muscle contraction work?'

The complexity of biological systems requires specific questions at each level of organization. Asking a good question is surely more than half the battle of doing good research. I am afraid it is difficult to resist using the opportunity of writing an introduction like this, to pontificate on one's own philosophy of research. I believe that the most difficult balance to hold in a scientist's division of labour is between the acquisition of broad vistas and concentration on a specialized and specific question at any one moment in time. No single series of experiments can tell one how rhodopsin transmits information about the absorption of a photon to the brain. However, the specialized experiments which one carries out with rhodopsin are only likely to contribute to our knowledge about visual perception if they are planned in the context of many other aspects of this field of endeavour. The problem with technique oriented biophysics is that, while it often concentrates on a particular question with each system, it does not permit sufficiently wide acquaintance with the behaviour of the system.

The above comments must not be taken to mean that I believe there is no other way. It merely points to the fact that in spite of the exciting results obtained by crystallographers or NMR spectroscopists, for instance, they have not solved the problem of 'how enzymes work'. Ideally one should ask a question and then look for a technique which can be used to answer it, rather than have a technique and look for a question it can answer. It is with this in mind that the publishers have asked a number of potential authors to write on biophysical questions, highlighting problems and surveying the range of approaches towards obtaining solutions for them. It is difficult to be sure at this stage which ones of those invited will produce a manuscript. The series will produce some volumes which attempt to answer questions about problems which have a wide application to the behaviour of biological systems. While this may appear to contradict my emphasis on concentration on one system at the time, such volumes will underline the importance of using a wide range of approaches to study those properties of molecules and their interactions which give information about the 'physical laws which govern the behaviour of biological systems'. One of the most important of these problems is the way macromolecules, like proteins and nucleic acids, behave in the 'real' concentrated solutions containing many other small ions which constitute their natural environment. The present volume presents a serious effort to treat this aspect of molecular biophysics. Other volumes, which are in an advanced state of preparation, will be concerned with the thermodynamics of ligand binding and other biochemical equilibria (Edsall and Gutfreund) and with the physical and chemical mechanisms of the action of calcium ions in their ubiquitous physiological functions (Campbell).

H. GUTFREUND

Contents

A Note on Symbols xv

Introduction 1

Chapter 1 Setting the scene **7**

1.1 Aqueous solutions........ 7

The iceberg concept and hydrophobicity; Practical applications to macromolecules

1.2 Thermodynamic implications 13

1.2.1 Gibbs–Duhem relationship........ 13

1.2.2 Osmotic pressure........ 14

Donnan potentials

1.2.3 Virial coefficients........ 18

B_2 for solutions; Polyelectrolytes

1.3 Chemical equilibrium........ 23

1.3.1 Energetics of equilibrium........ 24

1.3.2 Activity coefficient........ 25

Activities of electrolytes

1.3.3 Summarizing equilibrium 27

1.4 Molecular forces and energy........ 28

1.4.1 Ionic forces........ 29

Macromolecular implications

1.4.2 Hydrogen bond........ 31

1.4.3 Dipole interactions and van der Waals' forces 32

Energy and polarizability; Energies of molecules in solution

1.5 Conclusions 38

Chapter 2 Weighing molecules........ **39**

Historical trends: Fragility of macromolecules

2.1 Survey of methods for estimating relative molecular masses 42

2.1.1 Osmotic pressure........ 43

Non-ideality of haemoglobin solutions

2.1.2 Intensity light scattering 45

Elastic scattering; Non-ideality in light scattering

2.1.3 Ultracentrifuge.. 49
Velocity ultracentrifugation; Concentration dependence of $S_{20,w}$; Molecular weights estimated from the Svedberg equation; Sedimentation equilibrium; Analysing experimental records; Experimental considerations

2.2 Interpreting ultracentrifugal results........................... 60

2.2.1 Non-ideality 61

2.2.2 Solution densities 62
Pressure

2.2.3 Baselines.. 66
Refractometric measurements; Light absorption

2.3 Relative matrix methods 68

2.3.1 Gel chromatography 69

2.3.2 Gel electrophoresis.. 73

2.4 Applications.. 75

2.4.1 Heterogeneity and accuracy 75
Relative mass of albumin; Quantifying heterogeneity; Sensitivity of equilibrium methods to heterogeneity; Moment analysis for heterogeneity; Comparative techniques

2.4.2 Effect of baselines 85

2.5 Estimating relative molecular masses of large aggregates 87
Light scattering from large particles

2.6 Lessons to be learnt 90
(1) *Purity of sample;* (2) *Interaction with small diffusible ions;* (3) *Concentration ranges;* (4) *Calibration of the method;* (5) *Differential methods*

2.7 Appendix to Chapter 2 93

2.7.1 Derivation of the Svedberg equation........................ 93

2.7.2 Partial specific volumes of selected amino acids 95

Chapter 3 From friction to flexing.. 96

3.1 Defining shapes.. 97

3.1.1 Which technique is best?.. 98

3.1.2 What are hydrodynamic shapes?.. 98

3.2 Frictional coefficients.. 100

3.2.1 Known geometrical shapes 100
Translational movement; Rotational movement

3.2.2 Frictional coefficients for irregular geometrical shapes 101

3.2.3 Flexing of macromolecules 102
Relaxation times; Values of relaxation times; Harmonic oscillator model

3.3 Experimental parameters.. 106

3.3.1 Diffusion and friction.. 106
Effects of solvation

3.3.2 Concentration dependence of diffusion coefficients 108
Diffusion and osmosis; Intermolecular interactions; Theoretical studies of concentration dependence; Summarizing non-ideality and concentration dependence; Macromolecular association and diffusion

3.4 Estimating diffusional parameters 123
3.4.1 Diffusion by mass transfer 123
3.4.2 Quasi-elastic light scattering (QEL) 126
Interference equations; Autocorrelation; Correlations and diffusion; Diffusion and polydispersity; Large particles; Experimental; Temporal autocorrelation; Other correlation procedures; Raman scattering
3.4.3 Dielectric relaxation and other methods for measuring rotary diffusion coefficients 140
Rotation of serum albumin; Laser excitation; Fluorescent lifetimes

3.5 Lessons to be learnt from studying molecular motion 147
3.6 Appendix to Chapter 3 148
3.6.1 Relationship between osmotic pressure and diffusion 148
3.6.2 Frames of reference for kinematic viscosity (diffusion) 149
Dynamic viscosity

Chapter 4 Sticky molecules **151**
4.1 Basic mathematical models of specific macromolecular interactions 153
4.2 Estimating interaction parameters 155
4.2.1 Averaged parameters 155
Errors in $L_{1,2}$; Association in sedimentation velocity; Association and diffusion
4.2.2 Boundary analysis of associating systems 162
4.2.3 Equilibrium centrifugation 164
4.2.4 Qualitative tests for association 166
4.3 Association of biological macromolecules 168
4.3.1 Strong interactions 168
Lactate dehydrogenase; Haemoglobin
4.3.2 Medium interactions 173
4.3.3 Weak interactions 181
4.4 Dynamic relaxation 183
4.5 Conclusions 188

References **189**

Index **197**

A note on symbols

$\mathcal{N}$ Avogadro's number (number of molecules in one mole).

In general, all other symbols are defined as they appear but, in order to maintain usage according to convention for some symbols, confusion is possible between chapters. An attempt has been made to maintain conformity throughout and the exceptions are listed below.

f refers to activity coefficient in Chapter 1 and occasionally in Chapter 2; this symbol is also used to describe the molecular frictional coefficient in Chapters 2 and 3.

M_2 the subscript 2 refers to the property of the macromolecule in a ternary mixture containing (1) solvent, (2) macromolecule and (3) buffer ions; in Chapter 4, it refers to the property of the dimer.

Introduction

Experience over years of teaching and research has shown that one major challenge to workers studying the behaviour of macromolecules is to provide and utilize adequate explanations of the concentration dependence of their hydrodynamic and thermodynamic properties. Undergraduates and researchers are always pointing out that most thermodynamic and hydrodynamic concepts are best applied when working at infinite dilution and this is hardly a realistic experimental condition. Deviations between theory and experiment caused by working at finite concentrations are corrected through empirical or, at their best, semi-empirical coefficients. This approach suggests that these deviations are an annoyance which must be tolerated but are not included in theories. Possibly a more realistic approach would be to enquire whether these deviations from ideality, which are small for globular proteins, carry useful information about the state of the molecule in solution and, if so, is this reflected by biological properties of the macromolecule?

The standard way of dealing with these deviations is through virial coefficients, but transforming these into molecular properties is only covered adequately for neutral molecules using the concept of excluded volume. Furthermore, the coefficients are only suitable for discussing thermodynamic non-ideality but not hydrodynamic non-ideality, so they cannot be applied to discussion of concentration dependence of sedimentation coefficients nor to describe the complete deviation of diffusion coefficients. In many respects, the behaviour of macromolecules is very dependent on concentration, and this means that the full understanding of the deviations is necessary for two applications: (*a*) the exact correction of experimental observations on macromolecules when working at normal concentrations can only be made by allowing for the relatively large excluded volume of these molecules compared with, say, sucrose; and (*b*) a full understanding of their physico-chemical behaviour at finite concentrations must include concentration dependence if these data are to be used for descriptions of biological activity. Of these two reasons, the second is possibly the most fundamental, since it seems likely that the forces that cause the concentration dependence will be reflected in the biological activity. Even for this early part of the book, it seems necessary to justify this statement. One of the major problems in biology is to describe detailed processes by which cells or molecules recognize likeness (or foreignness): macroscopic experiences of this activity are many and range in complexity from rejection of grafts or organs in medicine through specificity of antigen–antibody reactions to purification by crystallization. Qualitative statements about recognition are difficult to extend down to the molecular level. Furthermore, since the expressions of molecular recognition normally occur through non-covalent interactions, it seems reasonable to explore the weaker physical forces between molecules. Thus, in solutions, this means understanding how the detailed structural features of the molecule modify the organization and activity of the solvent, as the frame of reference moves from the solute molecule out into the bulk solvent. Explanations of solute–solvent interactions are generally based on relative energies, and present trends are towards using these energies to support models given at the molecular level. This movement from rigorous thermodynamic interpretation of macroscopic observations to views more concerned with describing molecular motions is a continuing development in biochemistry. Studying diffusional processes at the molecular level is one part of this development. It was with these views in mind that this book was written.

Many authoritative texts are available which present accounts of the basic theories describing the solution properties of macromolecules; thus, detailed theory is often not presented in this text. Of the many books which could form 'necessary reading' are the one by Tanford (1961) and the recent one by Marshall (1978). One disappointing aspect of textbooks for students of all levels is the division that seems to exist between giving detailed descriptions of theories and outlining the flaws in experimental work which lead to frustration when testing or implementing theory. For this reason, I have tried to include examples of situations where inexperience or insufficiencies in the theory have led to false interpretations; it should be added that later work has sorted out these problems, otherwise we would not know that the original ideas carried flaws (normally the experimental data are reliable). These examples are obviously limited to problems that I have met in my own research, and consequently are not a full coverage for all techniques, but I hope the message conveyed is clear—always assume Murphy's law applies to experiments, i.e. that if results can be ambiguous or fogged by uncertainties, then this will occur!

Assuming experiments do give reliable results on the activities of macromolecules in solution, then what use is this to describing biological activities; in other words, what is the overall message of this book? Basically, it is directed to two categories of scientist: to biochemists, the message is not to underestimate the role of long-range molecular forces when describing the role of macromolecules in biology; and to biophysical chemists, it is to continue to provide exact models for molecular interpretations of concentration dependence. It became clear while compiling this book that a frontal attack on these forces and their relationship to concentration dependence was not a useful approach. Their discussion must be tied to the theoretical and experimental descriptions of the actual parameters involved. For this reason, these are discussed at some length. In fact, fewer words are devoted to discussing concentration dependence than to painting the background upon which the effects work. Nevertheless, the evidence shows that forces between macromolecules range over much larger distances than often thought; this evidence comes from the consideration of the dependences of sedimentation and diffusion coefficients. It is well known that the energies are weak (but may be larger than thermal energy), but it is less well known that the drop in energy with distance between macromolecules is relatively gradual. These may be interesting observations to physical chemists, but it is reasonable for biochemists to question whether this knowledge is important to the biological activity of macromolecules.

In answering this question, it is necessary to recall one fact that continually impresses physical scientists, namely the small range of temperatures that a biological system can tolerate. In itself, this observation can suggest either that the system is finely tuned for activity or that weak energies play important roles in regulating the system: both would mean that small changes in energy supplied to the system would produce effects far outside normal physical experiences. Observations show that radical shifts in the chemical status of natural macromolecules are produced by relatively small energy differences. For example, tetrameric haemoglobin readily dissociates into two dimers by dilution, while deoxyhaemoglobin only dissociates marginally even in excessively low dilutions, yet the overall difference in energy between the two contrasting states is only about 20 kJ mol^{-1} (or about eight times thermal energy at normal temperatures). Other equally surprising effects of low energy differences are commonplace with biological macromolecules, and Blow (1974) has suggested that energies that are considered reasonable are about 20 kJ mol^{-1} whereas those that are strong are about 80 kJ mol^{-1} (this reference was to the relatively high energy of interaction between trypsin and antitrypsin agents).

These lines of reasoning can be extended to structural work where large biological changes are often reflected by marginal changes in molecular structure: Perutz (1975) found that the structure of a thermally stable version of haemoglobin had only ten residues different in the β-chain from the common haemoglobin A. This change induced a few extra non-polar interactions which in turn produced no more than 5 kJ mol^{-1} extra energy for the thermostable haemoglobin. These findings all indicate that interactions through forces of relatively low energies are biologically important, and the concentration dependence of hydrodynamic parameters may be one way of measuring some of these low interactive energies. In this respect, it is interesting to find that macromolecular structures determined by crystallography are in themselves not leading now to many new concepts of activity of natural macromolecules (summarized by Mathews, 1977). Instead, attempts are being made to relate these structures to proportions of the surface exposed to the solvent (Lee and Richards, 1971; Chotia, 1974; Wodak and Janin, 1978) in attempts to incorporate concepts such as hydrophobicity (Kauzmann, 1959) into the structure. Does this mean that hydrodynamic and crystallographic studies are converging through the acceptance that a major factor in determining biological activity is the long-range weak forces of solute–solvent–solute interactions? Certainly, Anderson (1976a, b) would agree with this view, since he suggested that recognition of biological foreignness (e.g. an antibody recognizing an antigen molecule in a sea of other proteins) occurs through paratactic (or complementary) surface interactions, so pointing to surface tension as a determining influence (surface tension is a relatively weak force produced by disturbed solvent–solvent interactions at a phase interface).

Attempts to measure the interactions between macromolecules in solution have in the past been made through macroscopic observations. Then, using rigorous thermodynamic theory, attempts have been made to extrapolate down to the molecular level. This path is difficult and the models far from unique; hence, the move to measuring molecular motions through more direct procedures. One of these procedures involves the measurement and molecular interpretation of diffusion using the technique of quasi-elastic light scattering. With this method, it is possible to measure diffusion coefficients relatively accurately and quickly, so making it possible to study routinely the effects that small changes in molecular structure have on the motions of the molecules. Absolute values of the coefficients are possibly less important than the differences produced by varying treatments or changing concentrations. It is this latter phenomenon that provides hitherto untapped source of information for measuring the weak energies of interaction between molecules in solution. The complex theories that relate molecular models to the measurements are at present limited and are unable to cover many of the real situations pertinent to biology; nevertheless, these will improve. In fact, the situation could be likened to X-ray diffraction, which, 20 years ago, was an unknown quantity to most biochemists, but is now accepted without question, even at the undergraduate level. It is for these reasons that the discussions on diffusion (Chapter 3) are longer than would normally be expected and the description of quasi-elastic light scattering is given in some detail.

Only the future will decide whether these conjectures are borne out by experimental evidence, but it is a fact that most biologically active proteins are formed from oligomeric complexes containing several individual chains in the complex. The forces holding these chains together vary in strength from strong (but less strong than covalent forces) to relatively weak, and the view that even the weaker forces that occur between the oligomeric macromolecules moving free in solution is really an extension of these forces.

Summary

An attempt has been made to present descriptions of physical ideas with the minimum of mathematical derivaiton. With the exceptions of Sections 1.4.3, 3.3.2 and 3.4.2, only a moderate knowledge of mathematics and physics is required to appreciate the concepts. The three sections mentioned could be omitted from the first reading, but in any case a full understanding of the material in these sections will need supplementing through the reviews and books quoted in the text. The text contains sufficiently detailed references for workers to get the details if they so wish. This approach was chosen in order to present ideas to biochemists, who are familiar with handling proteins and nucleic acids and who can pose detailed biological questions, but who do not wish to survey the literature, which covers physico-chemical ideas, in order to examine these ideas. Thus, Chapter 1 provides a summary of many of the physico-chemical concepts used later in the book. Concepts concerning solute–H_2O interactions, virial coefficients, equilibrium constants, and origins of molecular forces are discussed at a variety of levels.

In Chapter 2 are given details of methods used to weigh macromolecules. The discussion covers experimental interpretations and precautions, concentration dependence of sedimentation coefficients, and relative mass. The problems raised by heterogeneity and solvation are discussed through results which were originally misinterpreted because of insufficient information.

In Chapter 3 is given a detailed discussion of macromolecular behaviour at the molecular level. This is followed by describing ideas that lead to discussion on solvated shapes and flexing of macromolecules. The main part of this chapter is concerned with translational and rotational movements of mass as measured by diffusion. It is through interpretations of these movements in a solvent that one can begin to recognize the importance of solute–solvent interaction on the behaviour of macromolecules. The experimental realization of these studies is through diffusion, and for this reason methods of measurement are discussed, paying particular attention to the role played by quasi-elastic light scattering in this work. Other methods such as dielectric relaxation are briefly discussed in order to maintain continuity with other texts.

Finally, Chapter 4 gives a survey of the specialized form of concentration dependence which can be described best through direct macromolecular association–dissociation reactions. The sources of errors and misinterpretations of earlier work are given for a few cases, together with a description of the influence that the strengths of interaction have on the measurements for proteins: these are particularly 'sticky' molecules.

I should finish this summary by stating that I have tried to avoid the propagation of misinterpretations, but I plead human frailty if I have committed the same sin of logic that befell a well known cartoon character. In the first picture of the series is shown a fluttering object in the distance and the character saying, 'Look at that butterfly, it has flown all the way from Brazil'. The next picture showed the companion examining the object and saying, 'This is no butterfly, it is a leaf'. The final picture showed the character saying, 'How did a leaf fly all the way from Brazil?' I feel sure most readers will agree that it is very easy to be funnelled into this type of argument when details obscure the true facts.

Acknowledgements

Extensive talks and discussions with many friends have been incorporated in the text and it would be difficult to list all these people. All I can do is to thank them and apologize to them if they feel I have incorporated their ideas in this text without due recognition: the subconscious is a powerful but forgetful force. There are, however, five people whom I want to thank specially: Professor G. A. Gilbert and Mrs L. Gilbert, who have suffered my inadequacies stoically over many years as well as acting as very intelligent sounding boards; Dr N. G. Anderson, for providing persuasive logic to clear many involved biological concepts; Dr J. M. Creeth, for carefully reading and criticizing this work—he offered many valuable suggestions and guided my reasoning in many places; finally, Professor H. Gutfreund, who has criticized, cajoled, and encouraged me to put some ideas down on paper. I have to thank NATO for the generous financial assistance which allowed me to work with Dr N. G. Anderson while my thoughts on solute interactions were related to biological problems. I also wish to thank the editors and publishers of the following journals for letting me incorporate various figures in the text: *Analytical Biochemistry*, *Biochemistry*, *Biophysical Journal*, *Biophysical Chemistry*, *Biopolymers*, *Chemical Society*, *Journal of Biological Chemistry*, *Journal of Molecular Biology*, *Journal of Physical Chemistry*, *Methods in Enzymology*, and *Nature*, *New Biology*.

CHAPTER 1

Setting the scene

Biological activity of macromolecules is normally associated with a solvated molecule either in suspension or in true solution. Thus, if one is to describe biological activity in physico-chemical terms, the interaction of biological macromolecules with their aqueous environment becomes a major determinant. The steps between carrying out an experiment and realizing from the results useful information on biological activity are often long and tortuous, involving the interpretation of complex interactions between all the molecules; sometimes it appears that the details cloud the overall picture. If one is to maintain clarity, then a good starting point would be from a discussion of the interactions of polyatomic molecules in solution. A full and detailed description of these interactions cannot be made for most molecules because this would involve complete formulation of wave equations using the rules of quantum mechanics and statistical thermodynamics. At their best, our physico-chemical descriptions tend to stop some distance from this and theoretical discussions falter after interpreting the macroscopic thermodynamic function. The strength (and weakness) of thermodynamics lies in its independence of molecular models, and its rigorous application for interpreting macroscopic results is inarguable. While accepting this limitation, from thermodynamics have flowed many useful models which provide interpretations of the dynamic activities of molecules; these have been constructed from solid models using coordinates determined by X-ray crystallography. These models rarely allow extrapolation to the quantized energy levels measured by spectroscopy. This chapter cannot do justice to the wide range of detailed theory which links spectroscopy with thermodynamics or X-ray crystallography with biological activity, but it attempts to summarize some of the components which are relevant to understanding the interactions between a macromolecular solute with the excess solvent molecules.

1.1 Aqueous solutions

First, one must consider the physical characteristics of water. Natural macromolecules are dissolved in water in concentrations which, in the biological cell, span several orders of magnitude. Many enzymes are present in relatively low average concentrations and these can be treated as very dilute ideal solutions, but others are present in concentrations which are far removed from ideal (for example, haemoglobin at 30–40% w/v, and muscle filaments and DNA in the nucleus where it is doubtful whether the term 'solution' can be applied). Throughout this range of concentrations, the introduction of the macromolecules into water produces marked changes in its macroscopic properties. Thermodynamic arguments show that these changes are reflected in the activities of the solutes. Normally, the liquid phase of water is considered to consist of monomers and polymers interacting through the well known hydrogen bond. Hydrogen bonding raises the boiling and melting points of water when compared with similar compounds formed from elements in the same group of the periodic table (e.g. H_2S). Furthermore, the fact that the size of the water polymers is not constant, but is a constantly changing dynamic equilibrium, means that any description of water must be through statistical averages.

The iceberg concept and hydrophobicity

Our basic ideas on the properties of aqueous solutions come from studying the effect that small solutes have on the properties of water. For example, introducing into water a solute which is not capable of forming hydrogen bonds (e.g. CH_4) excludes water from the volume occupied by the solute. This induces shells of organized water molecules arranged concentrically around the solute: the number of shells is ill defined, but rehydration studies suggest the existence of primary, secondary, tertiary, and more shells around all natural macromolecules dissolved in water (Texter, 1978). In colloid chemistry, these well defined lower layers are known as the Stern layer and the more diffuse layers the Gouy layers. Thermodynamically, this would be considered an entropy-based contribution to the energy, because the number of degrees of freedom of water in the region of methane molecules has been reduced. The effect is to produce a region where spatial reorganization of the hydrogen bonds of water is forbidden because of the presence of the methane. Instead, bonding can only occur around the surface of the methane to produce a relatively stable shell of water arranged in a configuration similar to that found in some forms of ice. Restricting the movement of water in this region profoundly changes its properties, producing an increased dielectric constant than in the bulk phase. This simple picture of the influence that a neutral molecule has on the structure of water was first proposed by Frank and Evans (1945) and, despite its age, the simplicity of the model has attracted many theorists who have used it to describe the activity of aqueous solutions. Basically, it provides an explanation of the negative enthalpy and entropy that occurs when a paraffin is mixed with water. Further, it fits the observation that dissolving paraffins in water is accompanied by a contraction of volume, which Nemethy and Scheraga (1962) pointed out was consistent with packing water in the large vacancies formed in ice under low external pressures (several well defined ice structures can be formed by varying the external pressures during crystallization). Simple extensions of this model to include solutions of biological macromolecules cannot be explored in detail here, but it is possible to discuss the factors and conclusions which follow from the model.

Possibly the most important concept to arise from the entropy effects of adding a paraffin to water is the so-called hydrophobic 'bond' of Kauzmann (1959). This is not a bond in the accepted sense of the word but more a contribution to the configurational energy of a mixed polymer in water (Tanford, 1978). It defines the tendency for units of the polymer to move into or out of the regions containing solvent moleculs. This energy is described by the laws of thermodynamics, giving the well known expression for the Gibbs energy ($\Delta\bar{G}$) of a system

$$\Delta\bar{G}_i = \Delta\bar{H}_i - T\Delta\bar{S}_i \qquad 1.1$$

where, for a mixture, the Gibbs energy, the enthalpy ($\Delta\bar{H}_i$), and the entropy ($\Delta\bar{S}_i$) are partial molar functions (denoted by the bars) in order to satisfy the Gibbs–Duhem relationship (Section 1.2.1). Thus, if $\Delta\bar{H}_i$ and $\Delta\bar{S}_i$ are both negative, the sign of $\Delta\bar{G}_i$ will depend on the magnitude of $T\Delta\bar{S}_i$. If a configurational change occurs during mixing or during exchange of solute between two solvents (or phases), the $\Delta\bar{S}_i$ of mixing equals the entropy of this configurational change. This, in turn, equals

$$k\ln(N_1 + N_2)!/N_1!N_2!$$

where N_1 and N_2 are the numbers of distinguishable states. In practice, N_1 and N_2 are unknown, so this description can only point to sources of an observed change in entropy, namely the total number of possible states in a given solvent. (For further details, see Moore, 1972; Tanford, 1961, Chapter 7.) If the substance is insoluble, then $\Delta\bar{G}_i$ will be positive, which means that $\Delta\bar{H}_i$must be small compared with $T\Delta\bar{S}_i$, inferring a significant configurational change ($\Delta\bar{H}_i$ for isothermal processes measures the exchange of energies produced by breaking or forming 'bonds'). When dissolving a solute in water without resulting chemical action (that is, unlike dissolving SO_3 in H_2O), $\Delta\bar{H}_i$ is equated with changed water–water and paraffin–paraffin interactions. The nett energy includes any differences produced by breaking and reforming hydrogen bonds as well as that arising from restructuring the molecule through rearrangements of van der Waals' forces between the paraffin chains. Among the many attempts made to rationalize these changes, the one that has attracted the most attention involves relating surface areas of the solute with its solubility. Hermann (1972) found that the logarithm of the solubility was linearly related to the area of aromatic molecules, and this was incorporated by Chotia (1974) in an empirical relationship to describe the solubility of amino acids in water. The conclusion of these workers was that about 100 J of hydrophobicity was associated with each square ångström of surface area—this is a free-energy change which favours removal of apolar surface areas from exposure to water. This source of energy is large and would correspond to between 1180 kJ mol^{-1} and 1480 kJ mol^{-1} for stabilizing lysozyme in solution (Creighton, 1978). Thus, the detailed energetics of dissolving a neutral molecule in water involve complex molecular interactions. Over a limited range of temperatures, the energies of these interactions are relatively constant (in fact, they vary as the exponential of T^{-1}), so the major influence that temperature exerts is on the entropy contributions ($T\Delta\bar{S}_i$, the configurational energy).

Tanford (1973) has considered in detail the thermodynamic implications of these solvation processes in relation to the hydrophobic effects. In order to discuss solvation, he first defines the effect that a dissolved hydrocarbon has on the chemical potential of water. To do this he needs to define the absolute chemical potentials of the solute in water, i.e.

$$\mu_w = \mu_w^\circ + RT \ln X_w + RT \ln f_w \tag{1.2}$$

where μ_w is the chemical potential of the solute in water containing a mole fraction of solute (X_w is the ratio of moles of water to the total number of moles of all the components in the mixture), while μ_w° is the standard chemical potential of the solute on the molal scale, i.e.

$$(\partial G/\partial n_w)_{T,P,n_2,n_3,\ldots} = \mu_w$$

where n_w is the number of moles of water. $RT \ln f_w$ is the excess chemical potential formed by solute–solute interactions (f_w is the activity coefficient); this becomes zero at infinite dilution (the reference state is infinite dilution of the solute). The

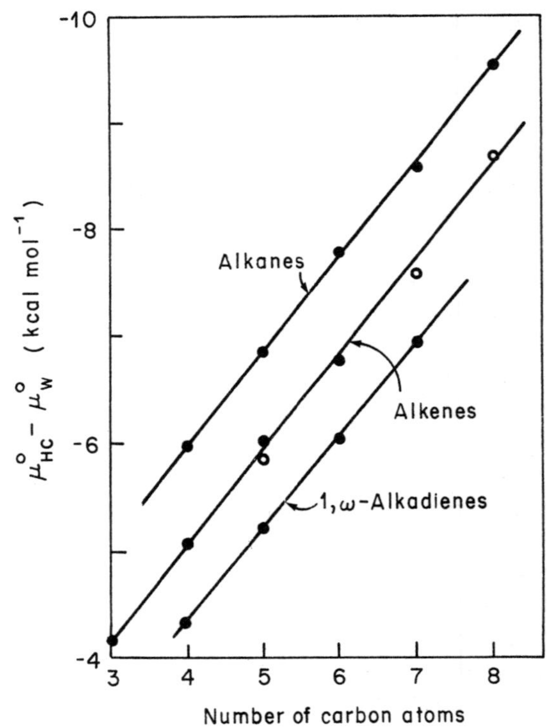

Figure 1.1 Standard free energies for transfer of hydrocarbons from water to pure liquid hydrocarbon at 25°C based on solubility measurements. For alkenes, filled circles represent 1-alkenes, and open circles represent 2-alkenes. (Taken from Tanford, 1973, Figure 2.1)

term $RT \ln X_w$ is necessary because of the statistical effects mentioned on p.9 and ascribed to the entropy of mixing the solvent and solute molecules. Similarly, the chemical potential of the solute in a hydrocarbon solvent (subscript HC) is

$$\mu_{HC} = \mu^{\circ}_{HC} + RT \ln X_{HC} + RT \ln f_{HC} \qquad 1.3$$

Assuming μ°_{HC} is similar for all hydrocarbon solvents, then the difference between the chemical potential of the two solvents describes the relative solubilities of the solute in hydrocarbon solvent compared with water. This difference is given as

$$\mu^{\circ}_{HC} - \mu^{\circ}_{w} = RT \ln (X_w/X_{HC}) + RT \ln (f_w/f_{HC}) \qquad 1.4$$

Figure 1.1 shows that increasing the carbon chain increases $(\mu^{\circ}_{HC} - \mu^{\circ}_{w})$ and provides a thermodynamic expression for Traube's rule describing the solubility of alcohols in water. The macroscopic corollary to the results shown in Figure 1.1 is that there occurs a decreased solubility in water with increasing chain length, a well established fact. These considerations led Nozaki and Tanford (1971) to propose a hydrophobicity scale based on the relative solubilities of amino acids in water and organic solvents (expressed as free-energy changes for the process). This scale is formed by subtracting the change in free energy of glycine from the changes in free energies of all the other amino acids and the scale ranged from -1.26 kJ mol^{-1} for serine to 1.43 kJ mol^{-1} for tryptophan. This scale could be interpreted as depicting the preferential probability of finding an amino acid out of contact with water when it is condensed in a protein.

Increasing the temperature of the mixture decreases the solubility of lyophobic

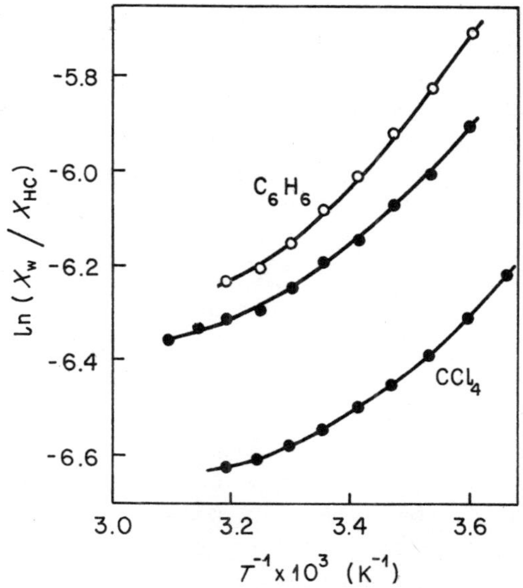

Figure 1.2 Logarithmic plot for the relative solubilities of ethane in water and in two organic solvents. (Taken from Tanford, 1973, Figure 4.1)

molecules in water. This experimental fact is interpreted as showing that the transfer of a solute from the hydrocarbon solvent to water is not energetically favoured. However, the relationship between ln (X_w/X_{HC}) and T^{-1} is often not linear (Figure 1.2). It would be expected that for simple ideal systems the relationship would be linear, so the non-linearity is caused by large differences in the heat capacities of the solute in water compared with that in the hydrocarbon solvent (the differences often exceed the individual absolute values for some systems (Tanford, 1973)). The differences that cause the non-linearity in Figure 1.2 cannot arise from the effect of temperature on the internal motion of the molecules, but arises from large structural changes in both solvent and solute caused by the introduction of the paraffin into a hydrogen-bonded system. In this process, bonds will be broken and the number of degrees of freedom of both water and solute will be temperature-dependent. It is these changes that are reflected by the changing heat capacities which produce the non-linearlity shown in Figure 1.2.

Practical applications to macromolecules

Transferring these arguments directly to the dissolution of natural macromolecules in water would suggest that macromolecules could only dissolve if hydrophobic groups could be folded out of contact with water. Natural limitations on the extent to which this folding can occur (restricted bond angles, etc.) show that hydrophobicity describes only part of the total energy of dissolving a macromolecule in water. The extent of folding that a polymer can undergo about its own centre of mass is limited by the amount that the covalent bonds can be distorted from their preferred orientations. Obviously, an energetic balance must occur in which energy derived from undesirable mixing of groups with solvent molecules must be balanced against opposing energies released when bending covalent bonds and breaking hydrogen bonds. The fact that hydrophobicity is only one of many factors which confer stability on a particular configuration is given by the finding that pressure-dependent precipitation of proteins is not accompanied by a change in sign of the relative molecular volume with increasing pressure (Zipp and Kauzmann, 1973). When a molecule is transferred from a hydrocarbon to water under increasing pressure, the sign of the relative volume should first be negative and then change to positive. For proteins, the sign remains negative over a wide range of pressures, showing that the positive contributions are too small to compensate for the negative volume changes even at high pressures (Brandts *et al.*, 1970). In other words, the configuration of the solvated molecule represents a stable minimum in the complex energy profile. This profile contains contributions from rearrangements of water molecules adjacent to the apolar residues which form a cavity. Evolutionary pressures will have decided whether this configuration has the desired biological activity, but often the conflict between insolubility and maintaining biological activity appears to be resolved through aggregation of subunits to produce oligomeric structures. In the contact areas, the hydrophobic groups are removed out of contact with water. Tanford (1978) emphasizes that this effect is of major importance in the organization of living matter.

Although it is necessary to build a qualitative picture of the interactions of solutes with water, there does not emerge a mathematical model which expresses quantitative relationships between individual macromolecules and water. It is only through the interpretation of thermodynamic functions for solutions that these qualitative models can be tested and their implications assessed.

1.2 Thermodynamic implications

Possibly the most obvious effect that solute–solvent interactions have on the properties of the solvent is in changing the surface tensions of liquids. Dissolving any solute in water either reduces or increases the surface tension; salts increase it, whereas many organic solutes (such as alcohols) reduce it. Thermodynamically, these changes can be described by the Gibbs adsorption isotherm (Moore, 1972, p. 484); it is also possible to relate this macroscopic phenomenon to molecular energetics. For example, when a lyophobic group is introduced into the solvent, it is possible to consider the void that it occupies as one phase, while the bulk water is the second. In this system, there is an ideal state which achieves equilibrium, making it possible to equate the energy required to maintain the interface with surface tension. This simple model cannot be discussed in great detail but it does provide a phenomenological approach to calculating lyophobic energies. Using this model, Kuntz and Kauzmann (1974) calculated the surface free energy of such a spherical void of radius 0.5 nm containing a paraffin group to be 94 kJ mol^{-1} of voids (the calculation assumes that the macroscopic surface tension of water can be transferred directly to molecular levels). This value is about 40 times greater than the disruptive thermal energy at 20°C ($RT \simeq 2.4$ kJ mol^{-1}). The excess energy would be sufficient to offset partially that adsorbed when the chains are folded to move the lyophobic groups out of contact with water molecules. It is at this point that the simple molecular picture begins to grow fuzzy, because we are now discussing energies rather than solid conformations. Macroscopically, these energies would be equated with thermodynamic energies, which in turn are discussed through chemical potentials.

1.2.1 Gibbs–Duhem relationship

For all mixtures, the dominating thermodynamic relationship is the one derived independently by Gibbs and Duhem. They showed that, for a mixture in equilibrium and at constant temperature and pressure, the following holds true

$$\sum_i X_i \mathrm{d}\mu_i = 0 \qquad 1.5$$

Here μ_i is the chemical potential of the ith component (and related to the Gibbs energy) which is present as mole fraction X_i. From eq. 1.5 it can be seen that adding components to a solvent reduces the chemical activity of the solvent ($\mathrm{d}\mu_i$

$= RT \ln a_i$, where a_i is the activity) by an amount determined by the mole fraction of the added component. This relationship is given by Raoult's law for partial pressures of ideal solutions by expressing the activity as a mole fraction of the solute, and this in turn relates the colligative properties to relative molecular masses, etc. (see Section 1.2.2). A more general view might be that eq. 1.5 stresses the importance of treating a solution as a whole system rather than a suspension of inert macromolecules in a continuous medium. Adding foreign molecules to a solvent changes both the solvent's chemical potential and the chemical potential of the solute; this follows from the earlier discussion of the hydrophobic effect in water p. 8). The extent to which this modifies the physico-chemical properties of the solution varies not only with the type of macromolecule but depends also on the type and quantity of the buffering ions in the solvent.

1.2.2 Osmotic pressure

From the Gibbs–Duhem relation (eq. 1.5) arise the colligative properties of solutions of which possibly the most basic macromolecular phenomenon is osmotic pressure. From this arises also the Donnan membrane potentials for charged macromolecules. Descriptions of osmotic pressure date back to Abbé Nollet in 1748, but possibly the first quantitative results were obtained by Pfeffer in 1877. Much later, the Earl of Berkeley and Hartley (1906) provided a sounder experimental basis for studying the phenomenon, while Adair (1925) was outstanding in applying the exact theories to biological macromolecules (discussed further in Chapter 2). Most physical chemistry texts give full discussions of the theory of osmotic pressure, but Tombs and Peacocke (1975) and Tanford (1961, p. 210) have gathered together the principles and practice of osmotic pressure as applied to biological macromolecules. Hence, it is only necessary to give here a brief summary of the relevant equations and their implications to macromolecule–solvent interactions.

Although the importance of osmotic pressure in measuring relative molecular masses of biological molecules has declined over the years, understanding the underlying physico-chemical principles is a great conceptual aid to future discussions. Very broadly, there are two groups of people in physical chemistry, those who restrict osmotic pressure to describing the phenomenon of a pressure generated across a semipermeable membrane separating solution from solvent and those who employ osmotic pressure to define chemical potential. For the purposes of this text, it is better to belong to the second category because it makes the relating of diffusion to osmotic pressure through differences in chemical potential (Chapter 3) easier. For similar reasons, it leads directly to discussions of differing chemical potentials across sedimenting boundaries where no semipermeable membrane exists but an external force is applied to oppose the broadening effects of diffusion (and equal to the gradient of osmotic pressure).

An osmotic pressure is generated when a solution is separated from its solvent by a barrier which is inpermeable to the solute but permeable to the solvent. Physically, this might be a membrane or an external force producing a gradient of concentration of the solute in a continuous solution. Under these conditions, Raoult's law for an ideal mixture states that the vapour pressure of the pure solvent (P_0) is reduced in proportion to the mole fraction of solute (X_2) in the mixture (where $X_1 + X_2 = 1$)

$$P = P_0(1 - X_2) \tag{1.6}$$

Mixtures that conform with this law are taken as being ideal by definition, and this occurs when there is no heat or volume changes occurring during dissolution of the solute. Guggenheim's work (1930) utilized this principle by assuming that, when the size of the neutral solvent and solute molecules where the same, the solutions were ideal. From this, Hildebrand and Scott (1950) developed the concept of regular solutions, where the intermolecular forces between solute and solute need not equal those between solvent and solvent molecules. Although these are important ideas, they are sufficiently well developed in physical chemistry textbooks to need no further discussion here.

When equilibrium is reached between the solvent on both sides of the barrier, the pressure on the solution side will have increased to P and its chemical potential will equal that of the pure solvent (μ_0). From conventional thermodynamic relationships, it follows that

$$\mu_0 = \mu + \int_{P_0}^{P} \left(\frac{\partial \mu}{\partial P}\right)_T \mathrm{d}P \tag{1.7}$$

The partial molar volumes of the solvent ($\bar{V}_1 = (\partial\mu/\partial P)_T$) and ($\mu_0 - \mu$) are related to the change in vapour pressure between the pure solvent and solution by

$$\mu_0(\text{vap}) - \mu(\text{vap}) = RT\ln(P_0/P) \tag{1.8}$$

For ideal solutions, $\mu_0(\text{vap}) = \mu_0$ and $\mu(\text{vap}) = \mu$ in eq. 1.7; therefore, assuming $\bar{V}_1$ is independent of pressure (that is, limiting the applicaiton to small differences in pressure), we have

$$RT\ln(P_0/P) = \int_{P_0}^{P} \bar{V}_1 \mathrm{d}P \tag{1.9}$$

$$= \bar{V}_1(P - P_0) \tag{1.10}$$

where $(P - P_0)$ is equal to the osmotic pressure of the solution (Π). It is now necessary to substitute for P_0/P in eq. 1.6 to give

$$-RT\ln(1 - X_2) = \Pi\bar{V}_1 \tag{1.11}$$

which reduces for dilute solutions (when $\ln(1 - X_2) = -X_2$) to

$$RTX_1 = \Pi\bar{V}_1 \tag{1.12}$$

X_2 may be replaced by $n_2/(n_1 + n_2)$ where n_2 equals the number of moles of solute contained in n_1 moles of solvent (relative molecular mass M_1). Theoretically, $\bar{V}_1$ should be measured before the molecular weight of the solute (M_2) can be calculated ($n_2 = w_2/M_2$ where w_2 are the number of grams of solute in n_1 moles of solvent). Also, $\bar{V}_1 = \mathrm{d}V/\mathrm{d}n_1$ (dV is the volume increase produced by introducing dn_1 solute molecules) and $\bar{V}_1$ can be determined by measuring the change in density (ρ) of the solution containing 1 mole of solute in V litres compared with the solvent (dρ/dc) for a range of concentrations, c, when

$$\rho = (n_1 M_1 + M_2)/1000V$$

and thus we obtain

$$\bar{V}_1 = \frac{\mathrm{d}V}{\mathrm{d}n_1} = \frac{M_1}{1000[\rho - c(\mathrm{d}\rho/\mathrm{d}c)]} \qquad 1.13$$

Incorporating eq. 1.13 into eq. 1.12 gives the general relationship used to determine relative molecular masses from osmotic pressures for an ideal solution, i.e.

$$RT\frac{w_2}{M_2}\frac{M_1}{w_1} = \frac{\Pi M_1}{1000[\rho - c(\mathrm{d}\rho/\mathrm{d}c)]} \qquad 1.14$$

However, the assumptions made in formulating eq. 1.14 restrict measurements to dilute solutions where $\rho_0 - c(\mathrm{d}\rho/\mathrm{d}c)$ is approximately equal to the density of the solvent (ρ_0) and $\bar{V}_1$ equals the molar volume of the pure solvent. In this case, $n_1 \bar{V}_1$ can be replaced by the volume of solvent associated with n_2 moles of solute (V') to give

$$\Pi V' = \frac{w_2}{M_2} RT \qquad 1.15$$

Replacing $w_2/M_2 V'$ by the concentration in moles per litre gives an equation analogous to van der Waals' equation for gases (Section 1.4.3), so making it possible to introduce virial coefficients when including non-ideality, thus

$$\Pi V' = \frac{w_2}{M_2} RT\left(1 + \frac{B_2(T)}{V^1} + \ldots\right) \qquad 1.16$$

Here $B_2(T)$ is phenomenologically related to $B_2'(T)$ of van der Waals' equation of state for a gas, but has different numerical values because of differing interaction energies for liquids compared with gases. $B_2(T)$ can be calculated from experimental records by plotting the reduced osmotic pressure (Π/w_2) against w_2. Any significant curvature in this plot reflects non-ideality, while the ideal osmotic pressure can be obtained by extrapolation to infinite dilution. It should be emphasized that these deviations are small for globular proteins and seldom measurable with any accuracy at concentrations below 10 g l^{-1}. This is not the case with expanded molecules (such as glycoproteins and nucleic acids) even when neutral, since here significant deviations from ideality will be observed at concentrations of 1 g l^{-1} (Creeth and Knight, 1968).

Donnan potentials

Two important conclusions can be drawn from the derivation of eq. 1.16. First, the osmotic pressure is derived from the changed chemical potential of the solvent caused by adding the solute (the discussion on aqueous solutions in Section 1.1 gives some insight into how this could happen at the molecular level) and, secondly, eq. 1.15 is derived assuming Raoult's law applies to the ideal solution. The implications of the second conclusion are considered in more detail in the discussions on ideality (p. 25). Equation 1.15 can be applied when solutions are sufficiently dilute that

$$\ln(1 - X_2) = -X_2$$

and for neutral molecules. Obviously, osmotic pressures are generated by charged molecules, and solutions of charged macromolecules carry counter-ions which may pass through the semipermeable membrane. This introduces the further constraint that, at equilibrium, both equality of chemical potential and electroneutrality must be maintained each side of the barrier. Because the polyampholyte cannot move across the membrane but the counter-ions can, an electrical potential difference is generated across the membrane, the Donnan effect. The practical and theoretical implications that arise from these two opposing effects lead to a more complicated formulation of osmotic pressure involving three equations (eqs. 1.17, derived by Rice and Nagasawa, 1961). These equations hold for a system containing a single salt as a counter-ion to a polyelectrolyte of charge q (as the anion in phase I) which at equilibrium produces an electromotive force (EMF) E across the membrane measured by electrodes reversible to the diffusible ion ($\bar{V}_1$ is the partial molar volume of the ith component):

$$qE + \bar{V}_{\mathrm{C}} + \Pi = RT \ln (X^{\mathrm{II}}_{\mathrm{C+}}/X^{\mathrm{I}}_{\mathrm{C+}}) \qquad 1.17a$$

$$-qE + \bar{V}_{\mathrm{A}} + \Pi = RT \ln (X^{\mathrm{II}}_{\mathrm{A'}}/X^{\mathrm{I}}_{\mathrm{A'}}) \qquad 1.17b$$

$$\bar{V}_{\mathrm{H_2O}}\Pi = RT \ln (X^{\mathrm{II}}_{\mathrm{H_2O}}/X^{\mathrm{I}}_{\mathrm{H_2O}}) \qquad 1.17c$$

The three equations contain three experimental variables (E, Π, and the amount of solute which diffused from solvent to solution), making it possible to calculate one uniquely knowing the other two. Normally, it is assumed the partial molar volumes are independent of pressure and the solutions are sufficiently dilute that

$$X^{\mathrm{I}}_{\mathrm{H_2O}} = X^{\mathrm{II}}_{\mathrm{H_2O}} = 1$$

which reduces the equations to

$$\Pi = \frac{RT}{\bar{V}_{\mathrm{H_2O}}} \ln\left(\frac{1 - X^{\mathrm{II}}_{\mathrm{C+}} - X^{\mathrm{II}}_{\mathrm{A'}}}{1 - X^{\mathrm{I}}_{\mathrm{C+}} - X^{\mathrm{I}}_{\mathrm{A'}} - X^{\mathrm{I}}_{\mathrm{P}}}\right) \qquad 1.18$$

($X^{\mathrm{I}}_{\mathrm{P}}$ is the mole fraction of the polyelectrolyte in phase I.) Expanding the logarithm and replacing the X's by weight concentrations in the manner used when forming eq. 1.14 from eq. 1.12 gives (see Rice and Nagasawa, 1961)

$$\Pi = w_P RT\left(\frac{1}{M_2} + \frac{w_P}{4\beta^2 c_S} + \dots\right) \qquad 1.19$$

where c_S is the molar concentration of supporting ions. Comparing eq. 1.19 with eq. 1.16 shows that the second virial coefficient

$$B_2 \equiv (4\beta^2 c_S)^{-1}$$

($\beta = M_2/zq$, z being the valency, so β is the relative molecular mass per unit charge.) The validity of eq. 1.19 is supported qualitatively by experimental results (Pals and Hermans, 1952) when the expected linear dependence of B_2 upon c_S^{-1} is found to occur but the slopes are not equal to $(4\beta^2)^{-1}$. More detailed models which allow for interaction between adjacent charges for highly charged polyelectrolytes (for example, eq. 3.12, but see Nagasawa and Takahashi, 1972) show that there is a strong positive contribution to B_2 which is not included in eq. 1.18 and is derived from unequal distribution of the counter-ions and electrostatic charge around the macromolecules.

1.2.3 Virial coefficients

The discussion on osmotic pressure has introduced the concept of non-ideality through empiricism giving the virial coefficients. These coefficients form a series (as in eq. 1.16 and eq. 1.19) and, except for eq. 1.19, it is not obvious whether in general the coefficients have any physical significance for molecules in solution. Historically, the introduction of the coefficients became necessary when it was found that the gas equations did not explain the properties of gases over wide-ranging conditions. The simple equations were replaced by equations of state like eq. 1.20 where empiricism was enlisted to explain the lack of correspondence between pressure, volume, and temperature over all applied conditions (as, for example, when approaching liquefaction):

$$PV = RT\left(1 + \frac{B_2'(T)}{V} + \frac{C(T)}{V^2} + \dots\right) \qquad 1.20$$

More specific equations were derived to account in part for the deviations of experiment from theory, and of these the most famous was van der Waals' equation (eq. 1.21):

$$\left(P + \frac{a}{V^2}\right)(V - b) = RT \qquad 1.21$$

The coefficient a was introduced to allow for the mutual attraction of molecules in the gas; this attraction reduced the observed pressure below that expected for ideal conditions, leading eventually to liquefaction at high pressures. The coefficient b accounted for the molecules occupying a finite volume in the gas and for the reduced available space for the motion of the molecules (previous laws assumed point molecules having neither interaction with one another nor occupying any volume). In eq. 1.21 b is related to $B_2'(T)/V$, the second virial coefficient, for a gas having no mutual interactions. This historical treatment of gaseous properties shows that the second virial coefficient of a neutral gas is a function of the volume of the molecules; for spherical inert molecules

$$b = 2\pi \mathcal{N} r_0^3/3$$

where r_0 is van der Waals' radius, and $\mathcal{N}$ is Avogadro's number.

These simple views were empirical and insufficient to allow calculations of $B_2'(T)$ to be made from known molecular parameters (really $B_2'(T) = b - a/RT$) because they did not include functions which describe the mutual forces between the molecules. Later in this chapter (Section 1.4.3) are given equations for describing these forces (arising from dipole–dipole interactions), but here it will suffice to note that a more exact definition of $B_2'(T)$ would be

$$B_2'(T) = 2\pi \mathcal{N} \int_0^\infty (1 - e^{-E/kT}) r^2 \, dr \qquad 1.22$$

where the integral contains both energy (E) and volume ($r^2\,dr$). If it were possible to formulate exactly the sources of the energy E (or van der Waals' radius, but not both), then $B_2'(T)$ could be factorized into the various molecular energy levels and the solvent–solute energies calculated. The energy term would be small for neutral molecules and large for highly charged molecules, making the second virial coefficient an all-embracing correction for gases. It is possible to modify this approach by introducing higher-order coefficients to correct for the parts of non-ideality which cannot be included in excluded volume. In later chapters, mathematical models will be given describing structural interactions between macromolecules which could be calculated if $B_2'(T)$ could be measured accurately (future discussion of the second virial coefficient will assume the temperature dependence of the coefficient but the (T) will be omitted).

B_2 *for solutions*

Experiments show that it is possible to transfer these ideas from inert gases to solutions. Unfortunately, direct transfer between the two states of matter is not possible because there always remains the question of what is van der Waals' radius for a solvated macromolecule. Furthermore, it is not possible to assign energies of interaction between macromolecules, although these energies will appear in equations discussing interactions (Chapter 3). Nevertheless, it seems reasonable to assume that the second virial coefficient measured from osmotic pressure and other methods is a measure of the volume occupied by the solute. With electrolytes, this simple view must be modified (see Section 1.4 and eq. 1.19) to include interactions between charged particles. Thus, for charged molecules, the second virial coefficient is often broken into two components, the simple or ideal part reflecting the excluded volume (b_0), while the second part attempts to allow for electrical interactions (b_e) so $B_2 = f(b_0, b_e)$ (eq. 1.19). Another approach is to assume that the third and higher coefficients allow for the electrical and other less well defined interactions.

Exact derivations of the second virial coefficient for macromolecules through considerations of the differences in polarizability of a neutral molecule relative to the solvent are not possible (see Section 1.4.3 for a discussion of these polarizabilities). All that is possible is to relate B_2 to geometric volumes using eccentricities defined by the ratio of the semi-axis of ellipsoids (l and d with $p = l/d$). Ishihara and Hayashida (1951a, b) derived the following formula for both prolate ($p > 1$) and oblate ($p < 1$) ellipsoids:

$$B_2 = \frac{4\mathcal{N}V}{M^2}\left\{\frac{1}{4} + \frac{3}{16}\left(1 + \frac{\sin^{-1} G}{(1 - G^2)G}\right)\left[1 + \frac{(1 - G)^2}{2G}\ln\left(\frac{1 + G}{1 - G}\right)\right]\right\} \quad 1.23$$

where

$$G = \begin{cases} 1 - p^{-2} & \text{for} \quad p > 1 \\ 1 - p^2 & \text{for} \quad p < 1 \end{cases}$$

and where V is the volume of the ellipsoid in molecular dimensions, so $V\mathcal{N}M$ is approximately equal to the partial molar volume with $\mathcal{N}$ being Avogadro's number. From eq. 1.23, it is possible to calculate B_2 knowing the overall average dimensions of the macromolecule. Usually, B_2 estimated from experiments is not sufficiently accurate that the reverse calculation giving molecular dimensions in solution can be made. However, by choosing the correct experimental procedures, it is possible to calculate intermolecular interaction energies from B_2 (see Chapter 3).

The change in the activity of the solvent caused by adding finite amounts of solutes to a solvent reflect all solute–solvent, solute–solute, and solvent–solvent interactions. The increasing importance of these interactions as the concentration is increased is included in the virial equations by adding higher powers of concentration (eq. 1.16 and eq. 1.20), but the second virial coefficient is assumed to cover interactions which can be collectively called 'excluded volume'. This view provides a rational basis for using molar volume instead of molar (or molal) concentrations when discussing the activity of macromolecules (Florey, 1953). Employing volume fractions emphasizes the relatively large proportion of the total volume of a solution that is occupied by a macromolecule as compared with solutions of small molecules of similar molarity; the solvent is excluded from this volume. For this reason, it is desirable to employ volume fractions when studying the concentration dependence of macromolecular parameters.

As eq. 1.23 shows, the magnitude of the virial coefficients will depend on the effective volume of the solvated molecule. This is calculated by making allowance for the geometrical shape of the molecule (Tanford, 1961; van Holde, 1971). Since the dimensions of macromolecules are large, they have large molar volumes compared with the solvent molecules, which means that macromolecular solutions may be non-ideal even in relatively dilute solutions. For example, even in 10^{-4} M solutions, estimates of the relative molecular masses of many macromolecules in solution (Chapter 2) will be less than the true value. For this estimation, the virial equation for a homogeneous solute becomes

$$M_2 = M_2^{\text{app}}(1 + 2B_2c + 3B_3c^2 + \ldots) \qquad 1.24$$

where M_2 is the true and M_2^{app} the apparent relative molecular mass of the solute at concentration c(w/v) (the integer coefficients, 2, 3, etc., apply to centrifugal but not osmotic measurements (Fujita, 1961)). For neutral molecules, the higher-order coefficients (that is, all except B_2) are generally negligible at concentrations less than 10 g l^{-1}. If the molecule is charged, then higher-order terms become significant even at this concentration, but the coefficient which should be included is determined by the valency of the supporting ions (Yphantis and Roark, 1972). A monovalent counter-ion leads to positive B_2, zero B_3, and negative B_4. There is no simple rule for determining how many coefficients must be included in these analyses. Statistical tests are the most convenient guide, but the resulting coefficients may only be empirical and need not reflect the molecular interactions, especially if systematic errors are present in the records.

Polyelectrolytes

When self-association occurs between neutral molecules, then experimental values of B_2 may be zero or even negative. However, for polyelectrolytes, a zero, or smaller-than-expected, B_2 does not necessarily imply that association is occurring. Simple theory for polymers having zq independent charges (z being valency) has already been described for osmotic pressure (eq. 1.19) and this can be given as (Donnan and Guggenheim, 1932)

$$B_2 = \frac{10^3(zq)^2}{4M_2^2c_S} \qquad 1.25$$

Even with more advanced theories, where interactions between adjacent charges are allowed (for example, eq. 3.14, but others are summarized in Nagasawa and Takahashi, 1972), the proportionality between B_2 and c_S^{-1} is maintained. Thus, increasing the concentration of supporting electrolyte decreases the contribution from B_2, making it possible to reach a condition where no concentration dependence of the apparent molecular mass is recorded. Practically, this occurs when the macromolecule carries only a small nett charge ($B_2 \propto z^2$, eq. 1.25), and may be achieved for many proteins at about pH 7. But, for more highly charged macromolecules, it is necessary to shield the charges by dissolving in high concentrations of salt (greater than 1 M). High levels of charge appear to increase the hydrodynamic volume of the macromolecules over those found for similarly shaped neutral macromolecules. Quantitative models for these systems are based on the Debye–Hückel theory (leading to eqs. 1.29 and 1.31). This, in turn, is reflected by marked changes in the concentration dependence of unshielded macromolecules discussed in Chapters 2 and 3.

Sometimes, it is possible to enlist a condition where the long-range electrostatic effects are equal in magnitude but opposite in sign to the solute–solvent interactions through excluded volume (the theta condition; Florey, 1953). Under these conditions, not only does the second virial coefficient become zero but so do other hydrodynamic concentration coefficients. For example, when polyriboadenylic acid is dissolved in 1 M NaCl at neutral pH and 25°C, the second virial coefficient becomes zero (Figure 1.3, taken from Eisenberg and Felsenfeld, 1967). Increasing the temperature above 25°C causes B_2 to decrease to become a negative value, then eventually it passes through a minimum as the temperature is increased. Further, Eisenberg and Felsenfeld (1967) found that the theta temperature decreased with increasing salt concentration. At theta conditions, the true unperturbed structure of polyriboadenylic acid can be calculated from the experimental results without having to allow for long-range solvent–solute forces. This fortunate arrangement is not often found for proteins. From the results, Eisenberg and Felsenfeld (1967) calculated the radius of gyration of the acid, and related this to the stacking of nucleic acid bases; this stacking is a primary force in determining the three-dimensional state of nucleic acids in solution.

Physical models which explain fully the phenomenon of theta conditions are uncertain. It is possible, however, that the condition may be related to the fact that self-association between macromolecules is more likely to occur when charges are either not present or are well shielded. For example, if reversible

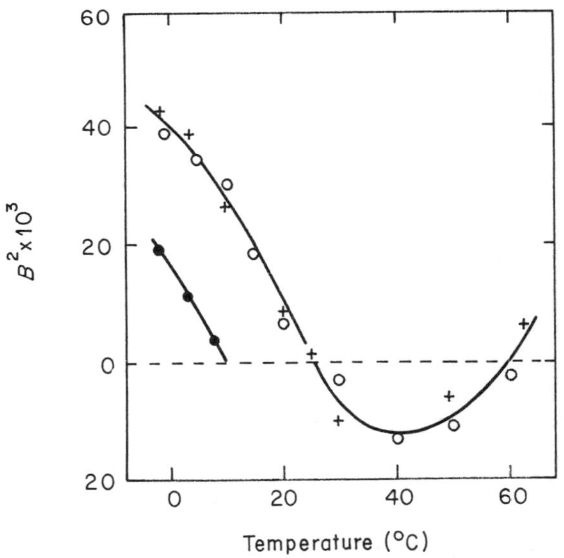

Figure 1.3 Second virial coefficient (B_2) of poly-A as a function of temperature (z is the relative charge, eq. 1.25): +-+-+-+, $z = 1740$, 1.0 M NaCl; o-o-o-o, $z = 1462$, 1.0 M NaCl; ●-●-●-●, $z = 1740$, 1.3 M NaCl. (Taken from Eisenberg and Felsenfeld, 1967, Figure 6. Reproduced with permission from *J. Molec. Biol.* **30** 17. Copyright by Academic Press Inc. (London) Ltd)

association does occur, then the average relative mass increases with increasing concentration (see Chapter 4), and if, if the association is weak, then the results of experiments could mirror those found under theta conditions. The approximate level of association necessary to offset a true positive B_2 for most proteins and yield zero concentration dependence would occur with a system having a dimerization coefficient of about $3\,\mathrm{l\,g^{-1}}$, for macromolecules having a charge of between 10 and 20 excess charges in 1 M salt. These levels of association are rarely reported with most proteins, but can occur under the special conditions employed in centrifuging lipoproteins when strong salt concentrations are required to increase the density of the solution: the addition of salt is necessary to increase the rate of sedimentation.

1.3 Chemical equilibrium

The concept of equilibrium is sufficiently well known that it seems unworthy of further discussion. Everyone knows of everyday situations where equilibrium exists. For example, a stone sitting in a depression on a hillside is at a metastable equilibrium, and if energy is expended in pushing the stone over the lip of the depression then the equilibrium is lost. Thus, an equilibrium is achieved when two equal and opposite forces exist. Doing work on the system adds energy which disturbs the equilibrium. The concept of equilibrium in chemistry underlies the laws of chemical thermodynamics, and for this reason the concept must be discussed further.

Macroscopic definitions of equilibria stem from the laws of mass action in which an equilibrium constant (K) can be described for simple reactions of the type

$$n\mathrm{A} + m\mathrm{B} + \ldots \underset{k_{-1}}{\overset{k_1}{\rightleftharpoons}} p\mathrm{C} + q\mathrm{D} + \ldots \qquad 1.26a$$

thus

$$K = \frac{C^p D^q \ldots}{A^n B^m \ldots} = \frac{k_1}{k_{-1}} \qquad 1.26b$$

For this definition to have any meaning, the symbols A, B, C, D, etc., must refer to chemical activities of the components; these are only equal to concentration at infinite dilution. This simple phenomenological definition does not convey much information about the chemical energetics of the reaction; it merely indicates the state produced when no net difference in chemical potential is recorded.

If one returns to energetic considerations, then it is apparent that, during a chemical reaction under adiabatic conditions, the initial total molecular energies of the components become distributed between the molecules of the unchanged reactants and those of the products at equilibrium. Thus, when the system reaches equilibrium, the total energy of an isolated system has not changed, but it has been redistributed (the criterion of conservation of energy). If the reaction was radiationless and isothermal, then the energy at each stage could be described from the rotational, translational, vibrational, and electronic states of the components. Seldom can this be achieved practically, and generally all that is available is a measurement of changes in concentration (or pressure) as well as the heat changes during the reaction. These are valuable measurements giving information about the state of the molecules, but from these alone it would be impossible to give anything approaching a full description of the reaction. More information on the states of the molecule can be obtained only from detailed spectroscopic studies. This concept of conversion of energy forms a useful basis for defining equilibrium.

1.3.1 Energetics of equilibrium

Balanced energetic equations form the basis of all theoretical discussions on chemical reactions, but calculating absolute energies of molecules before and after a reaction are seldom needed when discussing properties of macromolecules. What is required is the nett Gibbs energy change (ΔG°) occurring during the reaction, which for an isothermal system at constant pressure is defined as

$$\Delta G^\circ = -RT\ln K \qquad 1.27$$

This definition raises the problem that, in the absence of absolute definitions of energy, values can only be quoted with respect to some reference level. This problem can be illustrated by returning to the stone on the hill, where the potential energy at rest will be mgh (m is the mass, g is the gravitational acceleration, and h is the height) and the height is measured from a standard reference level on Earth where g is known. In this particular case, the energy would be described as an extensive property of the stone because it depends on the mass of the stone. However, its density would be an intensive property, since this is a specific characteristic of the material of the stone. These definitions can be transposed to all thermodynamic parameters, where other important intensive properties such as pressure and temperature become important. Thus, for the stone, one measures the potential-energy change with respect to the energy that the stone would have at a defined reference state called the standard state. The conditions of this state may vary according to the results being discussed, but are not universally absolute. Often in themodynamics it is defined as a pressure of 1 bar and temperature of 25°C, while ΔG° refers to changes of energy of all components at molar or molal concentrations. Studies on solubility are an example where a careful redefinition of the reference states provides a clearer picture of the underlying processes that determine solubility (see hydrophobic effects, Section 1.1, and Tanford, 1973). Gurney (1953), Kauzman (1959), and Tanford (1973) all discuss at some length the reference states used to give equations resembling eqs. 1.2 and 1.3. Here, if X is expressed as a mole fraction, then μ_w° and μ_{HC}° refer to infinite dilutions in water (eq. 1.2) or hydrocarbon (eq. 1.3). This means that all the interactions of the isolated hydrocarbon with water are included in μ_w°, and that the separate entropy contributions that arise from entropy changes produced by mixing are contained in $RT\ln X_w$ (eqs. 1.2

and 1.3). This leaves μ_w° to contain the internal energy of the molecule plus the energy of the interaction with the solvent.

1.3.2 Activity coefficient

It may seem that these discussions have digressed far from the opening phrases on equilibria, but this was intentional. The idea was to show that measuring thermodynamic parameters is only part of the process of defining the interaction between molecules, even at equilibrium. From all points of view, one needs to know how the definition given in eq. 1.26 is related to experimental variables measured at finite concentrations when activities are not generally available (electrode potentials and measurements of the apparent relative molecular mass in the centrifuge are the only experimental approaches that measure chemical activities of solutes directly). From the discussion on virial coefficients, it seems reasonable to assume that similar coefficients will be applied to other thermodynamic discussions concerned with intensive properties of real systems. Introducing a coefficient which experimentally describes deviations from ideality is common practice. This activity coefficient (f) is used as a multiplier for concentrations in order to make activity equal to fc. Thus, the Gibbs energy of eq. 1.27 becomes

$$\Delta G = -RT\ln\left(\frac{f_A c_A \cdot f_B c_B \cdot \ldots}{f_C c_C \cdot f_D c_D \cdot \ldots}\right) \qquad 1.28$$

where the concentrations can be molar, molal, or weight per unit volume (w/v). In all cases, the activity coefficient will remain a measure of non-ideality (it may seem to be heresy to quote a standard free energy using concentrations of $\mathrm{g\,l^{-1}}$ rather than molar, but logarithmic functions are dimensionless so, although the numerical values change with differing units of concentration, this will be accommodated in the definition of the reference state). Expanding the logarithmic function in eq. 1.28 produces an equation which resembles the equation of state for a gas (eq. 1.20).

From the definition of activity coefficient given by eq. 1.28, one would imagine that f has the limits $1 > f > 0$, but experience shows that often this is untrue. When dealing with solutions, the Gibbs–Duhem relationship comes to the aid of the unwary (eq. 1.5), showing that although

$$d\mu_i = RT\,d\ln a_i = RT\,d\ln fc_i$$

this definition is incomplete because there is more than one component in the solution. Thus, as f decreases with increasing concentration of one component, it increases for a second, and so on. These changes are not trivial, as can be seen from the examples given in Table 1.1 for a variety of simple salts. Even for NaCl, the activity coefficient goes through a minimum in the range 1 M to 1.6 M and then increases to approach unity at about 6 M. With lithium salts, a similar trend occurs. As the examples show, the extent of the increase gets larger with increasing anion size. These effects are more exaggerated with divalent cations

Table 1.1 Activity coefficients of aqueous solutions of various electrolytes at 25°C. (Data were selected from Tables 11 and 14 of the appendix of Robinson and Stokes, 1959, *Electrolyte Solutions*. Reproduced by permission of the authors.)

Molality	NaCl	LiI	ZnI_2	$Zn(ClO_4)_2$	$UO_2(ClO_4)_2$	LiCl
0.1	0.778	0.815	0.572	0.568	0.604	0.790
0.5	0.681	0.824	0.605	0.615	0.762	0.753
1.0	0.657	0.91	0.789	0.909	1.341	0.774
1.6	0.657	1.063	0.957	1.681	3.06	0.853
2.0	0.668	1.198	1.012	2.68	5.7	0.921
2.5	0.688	1.418	1.053	5.04	12.90	1.026
3.0	0.714	1.715	1.106	9.77	29.8	1.156
3.5	0.746	—	1.170	19.17	67.90	1.317
4.0	0.786	—	1.239	37.9	154.6	1.510
4.5	0.826	—	1.336	—	345.0	1.741
5.0	0.874	—	1.453	—	724.0	2.02
5.5	0.928	—	1.558	—	1457.0	2.34
6.0	0.986	—	1.747	—	—	2.72

(Zn^{2+}), and when the anion is particularly bulky (ClO_4^-) the activity coefficient rises to about 38 in 4 M solution; the example given for the uranyl salt is about the limit reported. These examples of the concentration dependence of activity coefficients must raise the question of why the values exceed the boundary values expected for f? To answer this, the activity coefficients of the solvent must also be measured, when it would be found that it decreased to negligible levels when the activity coefficient of the solute increased so that $f \gg 1$.

Activities of electrolytes

A quantitative model which describes the activities of ions in solution will obviously be based on the Debye–Hückel theory for ions in dilute solutions (Chu, 1967, Chapter 2; Gabler, 1978, Chapter 7). Historical considerations produce the following equation:

$$\ln f_{\pm} = -\frac{|z_1 z_2| e^2}{2\varepsilon k T} \frac{\kappa}{1 + \kappa r} \tag{1.29}$$

Here z_1 and z_2 are the valencies of the spherical ions of radius r in a medium having a dielectric constant of ε; κ is the reciprocal ionic radius defined as

$$\kappa = \left(\frac{8\pi \mathcal{N} e^2}{100 \varepsilon k T}\right)^{1/2} I^{1/2}$$

where I is the ionic strength

$$I \equiv \frac{1}{2} \sum_i z_i^2 M_i$$

In order to reduce the many limitations inherent in eq. 1.29 (for these, see

Robinson and Stokes, 1959), Fowler and Guggenheim (1949) derived a more exact relationship which differs from eq. 1.29 by including a term for the partial molal volume ($\bar{V}$, eq. 1.30), i.e.

$$\ln f_{\pm} = -\frac{|z_1 z_2| e^2 \kappa}{2\varepsilon k T(1+\kappa r)} + \frac{\bar{V}/v}{24\pi \mathcal{N} r^3}(\kappa r)^3 \sigma(\kappa r) \qquad 1.30$$

where v is the number of moles of ions formed from one mole of electrolyte, while

$$\sigma^2(\kappa r) = \frac{3}{\kappa r}\int_0^{\kappa r}\left(\frac{\kappa r}{1+\kappa r}\right)^2 \mathrm{d}(\kappa r)$$

For most ions, the right-hand term in eq. 1.30 has insignificant effect when κr is small, but increases sharply as this factor increases, as for example occurs with macromolecular ions. From eq. 1.30 can be derived the equivalent equation for the activity of the solvent (f_S) which also includes the partial molal volume of the solvent ($\bar{V}_S$) (Robinson and Stokes, 1959):

$$\ln f_S = \ln n_S + \frac{\bar{V}_S(\kappa r)^3 \sigma(\kappa r)}{24\pi \mathcal{N} r^3} \qquad 1.31$$

where n_S is the mole fraction of the solvent (for water containing 2 M solute this is about 24/25). Here, the numerical value of the second term may be small, but physically it accounts for the deviation from solvent ideality, and it is this small deviation which gives rise to osmosis and the colligative properties of solutions of electrolytes. So, when discussing these solutions, it cannot be dismissed as being negligible.

1.3.3 Summarizing equilibrium

The macroscopic definitions of equilibrium (eq. 1.26b) give an impression that any system at equilibrium is static and that no dynamic changes take place. This is far from the truth, as can be seen if the equilibrium constant is redefined as a ratio of forward to backward reaction rates (eq. 1.26b). It then becomes a temporal probability function defining the probability of finding, in any small volume of solution, p molecules of C, q of D, n of A, m of B, etc. (eq. 1.26a). If the time taken to sample the molecules in the volume element is considerably less than the reaction rates, then the proportion of the molecules recorded for each species will continually oscillate about a mean whose value will be the temporal average as defined by eq. 1.26b, or by ratios of the reaction rates. Sampling at times greater than the reaction rates will give the average value, and this is the normal situation when examining a system at equilibrium. These two sampling rates are factors which become important when studying the association–dissociation of macromolecular complexes using transport methods (see Chapter 4). From these arguments, it can be shown that, although equilibrium is a useful macroscopic state for thermodynamics, it has no place when considering microscopic states or molecular motions (as considered in Chapter 3). Even at absolute zero temperature, when random thermally induced fluctuations are zero, there still occur molecular motions derived from the motions of the electrons, called the zero-point energy of the system. Thus, any conclusions drawn from a macroscopic equilibrium can only lead to discussions of average energy balance or the proportions of molecules present in one of two states; from it, no information is given about dynamic aspects of the molecule. To get this information, experiments must be made which sample in a temporal scale less than the reaction rates of the fluctuation. Methods are available for doing this and some are discussed in Chapter 3.

1.4 Molecular forces and energy

In the previous sections, reference is made to energies of interaction between molecules. This terminology has its origin in everyday life, where continual references to work and effort are common. Normally, the energies are dissipated in self-evident ways, as Newton realized when he stated that every action has an opposing reaction. At the molecular level, the opposing causes are not so evident. A solution of macromolecules consists of a polymer in which atoms are joined by covalent bonds plus solvent molecules whose size is small compared with the whole polymer. Polymers interact with both the neighbouring polymers and the solvent molecules through forces which are considerably weaker than the covalent bond (a force equals $\mathrm{d}E/\mathrm{d}x$ where E is the energy and x is the distance, so discussions of energy and forces are interchangeable). This means that, although the polymer is considered to be a continuous entity in which bonds can only be broken by expenditure of considerable thermal or chemical energy, the low-energy interactions are more easily broken, leaving the chains in a continual state of flux. This provides a picture of a polymer in which the weak energies are those which have values less than about 240 kJ mol^{-1} of bond. In this discussion, covalent forces are not considered further because, although they determine the assembly of a macromolecule, often it is the weaker forces that determine the biological expression of activity of the polymer. It is useful to discuss the forces in more detail in order of decreasing energy but not necessarily in order of decreasing importance. Each type of bond plays its role in determining the conformation, and in some cases the concerted effect of a large number of weak forces at several points along a macromolecule can overwhelm the contribution from a few forces which individually have higher energy. The stability of biologically active proteins in solution depends on the balance between these forces and the effect leads to concepts such as the hydrophobic effect (Kauzmann, 1959; and discussed in Section 1.1). Even the generally regarded weak van der Waals' forces sum to relatively large energies per molecule when long chains are aligned together (as in polythene sheets). Detailed theoretical descriptions of these molecular forces are complex, and it is largely through the work of Debye that we can approach a thorough understanding of the quantitative aspects involved (this work of Debye is summarized by Chu, 1967). Physically, all these forces arise from alternating electrical fields (a useful summary of the interactions is given by Gabler, 1978), and consequently they should be treated as a single topic. However, for convenience, the origins of the forces are discussed in separate sections.

1.4.1 Ionic forces

These forces are often referred to as salt bridges (not to be confused with similarly named bridges employed in measuring electrode potentials), since they link groupings of opposite charge (for details, see Gabler, 1978, Chapter 2). As with all electrostatic forces, their magnitude is inversely proportional to the square of the distance (r) between the charges, i.e.

$$F = \frac{Lq_1 q_2}{r^2} R_1 \qquad 1.32$$

In eq. 1.32, q_i are the charges, L is a proportionality constant and R_1 is the unitvector ($+R$ implies repulsion, $-R$ implies attraction between the charges). Mathematically, eq. 1.32 resembles the law for gravitational attraction between bodies, but the magnitude of electrostatic forces is about 10^{25} times greater than the gravitational forces. The proportionality constant L equals unity in the c.g.s. system; when expressing F in newtons, r in metres, and q in coulombs (m.k.s. or SI system), $L = 8.9874 \times 10^9 \text{ N m}^2 3^{-2}$ ($L = 1/(4\pi\varepsilon_0)$ where ε_0 is the permittivity of free space). Comparisons of the energies of weak forces show that electrostatic forces are the strongest (the force between two charges each of 1 C at a distance of 1 m apart is about -9×10^9 N or 2×10^9 lb!). Similar calculations show that the force between two oppositely charged groups each carrying one unit charge and placed 0.2 nm apart is about -9×10^{-9} N; the equivalent angular velocity required for a laboratory centrifuge so that it equals this force is about $10^{16} \text{ rad s}^{-1}$, showing that gravitational forces on Earth cannot disrupt an ensemble of molecules aligned through charge interactions.

The relationship between a charged molecule and its counter-ions (and water) are dealt with through mathematical solutions of the Poisson equation, of which possibly the best known is that produced by Debye and Hückel (1923). The implications of this work on physical chemistry are well known, and its description is contained in all textbooks of physical chemistry, so they need not be repeated here. A survey of the application of the Debye–Hückel equation to biochemical systems is given in some detail by Gabler (1978, Chapter 7).

Macromolecular implications

Since the energy of ionic forces decreases inversely with distance, their effects extend out to molecules separated by relatively large distances. From a practical aspect, their interactive energies can be modified by changes in the surrounding environment. For example, when a polyampholyte is dissolved in water, changing the pH adjusts the proportion of acid to base groups that are charged, while adding salt increases the dielectric constant compared with pure water, leading to a decreased force between the groups. These environmental changes alter the interactive energy by orders of magnitude. Many examples exist where electrostatic forces are responsible for producing very stable structures. Coat proteins on viruses are often held to nucleic acids through electrostatic charges, making it difficult to separate the two. Normally, the separation of these two polymers *in vitro* can only be achieved by changing the pH; decreasing the pH decreases the number of ionized carboxyl groups, while raising the pH introduces a large number of new charges which repel one another and make the protein swell. A further procedure for separating charge-pairs is to add alcohols to the solution, so decreasing the dielectric constant compared with water. Unfortunately, this change is not the only effect that follows from adding alcohols to solutions of macromolecules, because apolar molecules penetrate regions of the protein which are inaccessible to water. Introducing the small molecule into these regions increases the volume of the macromolecules, and in oligomeric structures this can lever the subunit away from the complex. Even with single chains, the volume change produced by interpenetration can be considerable and may lead to confusion when interpreting experiments involving changing environmental conditions. For example, Chiancone and Gilbert (1965) give results for haemoglobin which could have been interpreted as a straightforward

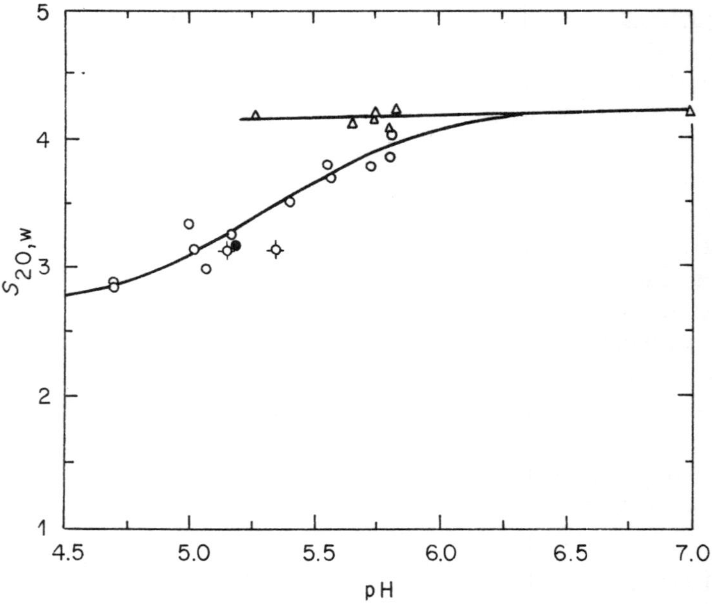

Figure 1.4 Relationship between $S_{20,w}$ and pH in two buffers for oxyhaemoglobin. The concentration of oxyhaemoglobin was 0.4% and ionic strength was 0.25 M. Buffers: ○, sodium acetate; ●, potassium acetate; -◇-, ammonium acetate; Δ, potassium phosphate. (Taken from Chiancone and Gilbert, 1965, *J. Biol. Chem.* **240** 3866, Figure 1. Reproduced by permission of the American Society of Biological Chemists, Inc)

change in ionic forces. The $S_{20,w}$ (see Section 2.1.3 for a definition of $S_{20,w}$) decreased markedly when the pH was reduced using acetate as the buffer (Figure1.4). It is known that salt bridges play a major role in the functioning and assembly of tetrameric haemoglobin, so it was expected that changing the pH would change the associative energies between the subunits. However, when the same pH range was studied using phosphate buffers (Figure 1.4), the change with decreasing pH was insignificant. These experiments show that ionic effects have been compounded with the penetration of undissociated acetic acid into the hydrophobic regions of the protein chains, and enchanced the dissociation as well as changed the shape of the molecule (this reduces $S_{20,w}$, see Chapter 2). These experiments show the care required when assigning causes for changes induced by comparing result from a range of solvents.

Studies with haemoglobin provide a further example of a situation where a chemical reaction induces a conformational change which leads to the formation of a new salt bridge between subunits. Deoxygenation of haemoglobin is accompanied by a considerable reduction in the dissociation of the subunits (see Chapter 4) and is accompanied by the formation of the new salt bridge. The changed ionic state of the interactions swamp the hydrophobic interactions. Normally, ionic forces between natural macromolecules are much greater than other non-covalent forces, but there are occasions when the resulting balance between the two sources of energy can produce stable aggregates of particles (see Section 1.4.3).

1.4.2 Hydrogen bond

Of the weaker forces between molecules, the hydrogen bond has the widest implications in biology (it is not a bond in the sense used for a covalent bond, but more a weak force—however, accepted terminology calls it a bond). This weak force is responsible for maintaining polar icecaps and keeps them floating on a sea which is still liquid at body temperature. It provides stabilizing forces for the ternary and quaternary structures of most biological macromolecules, as well as playing a vital role in many enzyme–substrate reactions. Dickerson and Geis (1969) provide extensive evidence of its importance to the assembly of biological systems. Despite its importance in stabilizing native macromolecules, an exact wave-mechanical treatment of the force remains elusive; the problems encountered in this treatment are best summarized by the statement of Ratner and Sabin (1973) that, 'Hydrogen bonds are *liaisons dangereuses* for theoreticians'. One of the earliest descriptions of the bond was made by Latimer and Rodebush (1920) who noted that when the hydrogen nucleus falls between selected atoms it is held by the two octets to form a weak bond.

One picture that has been drawn to portray the hydrogen bond comes from the Lowry–Brønsted acid–base theories, which show that a bond of the form A–H···B can occur between two atoms in which A is the proton donor and B is the acceptor. This qualitative description has been largely displaced by more exact treatments. The dissociation energy for this bond is small relative to a covalent bond or ionic forces (ranging between 20 kJ mol^{-1} to 40 kJ mol^{-1}), but large compared with thermal energy (about 2.4 kJ mol^{-1} at 20°C). Electronegative atoms of the first row of the periodic table (O, F, N) are commonly associated with the bond, but other elements can induce a bond (e.g. B). The presence of this preferred interaction is normally detected by perturbations in infrared, Raman, and NMR spectroscopy (for reviews, see Pimentel and McClellan, 1960, 1971; Murthy and Rao, 1968; Ratner and Sabin, 1973). Early theories were combined in Pauling's theory of the bond (Pauling, 1928, 1960) and give a view that since the A–H bond utilizes only 1 s orbitals of hydrogen, other interactions must be through point-charge electrostatics. This model, and others based on it, suffered from the drawback that the calculated bond strengths did not correlate with dipole moments of the molecules nor could they predict properly the changes in infrared intensity arising from the A–H stretching mode. Quantum-mechanical calculations based on valence-bond schemes (reviewed by Ratner and Sabin, 1973) show that the pure electrostatic picture is not satisfactory and that the 'covalent contribution' is important, particularly at short distances (less than 2.5 Å). The energy of the bond becomes large when it involves charged atoms, producing energies as high as 134 kJ mol^{-1} for $H_5O_2^+$ (a species formed in the upper atmosphere: Kraemer and Diercksen, 1970); this increase could be important in many ionized biological systems. A further factor of importance for calculating energies of the bond between biological molecules is the fact that the energy of two hydrogen bonds formed within the same molecule is not the sum of the two independent bond energies. Admittedly, this problem is more apparent when examining interactions between simple molecules (FH···FH···), but its implications should be considered when discussing interactions between a macromolecule and small molecules in solution or in G–C (guanine–cytosine) interactions of nucleic acids.

The outcome of these theoretical deliberations is to show that energetically the hydrogen bond is an ill-defined entity. Under some specialized circumstances, it possesses energies which approach those of covalent bonds, but generally its energy is low. The importance of the bond comes from its numbers in a macromolecule rather than from the properties of the individual bonds. The sum of the energies from bonds formed between adjacent groupings in a macromolecule can be large, endowing it with a particular conformational stability. Further, the strong interactions of biological macromolecules with water to form bound shells (for a review of the hydration of proteins, see Kuntz and Kauzmann, 1974) makes the forming and breaking of the bond a major source of energetic balance for reactions which involve the replacement of bound water on macromolecules (also see Pethig, 1979, Chapter 4).

1.4.3 Dipole interactions and van der Waals' forces

These weak forces are responsible for many of the macroscopic phenomena of everyday life, such as friction between surfaces and the strength of polythene sheets. The forces arise from alignment of the atomic/molecular dipoles, producing total interactive energies lying in the range from thermal, kT, to tens of kilojoules per mole. The energies of interaction between individual dipoles is always small and sometimes only exceeds kT when the temperature approaches absolute zero (as with He), but in polyatomic molecules these energies become additive to contribute a substantial part of the attractive forces. In biological systems, their influence is large, contributing a substantial part of hydrophobic phenomena (see Section 1.1) and producing stable arrangements of biological cells at fixed distances apart; however, the energy falls off as r^{-6}. This dependence suggests that they could normally be ignored when discussing the activity of macromolecules in solution, but often these forces form the only source of attractive energy between neutral molecules (e.g. polystyrene spheres and fats). Calculations of their contributions to the energy terms appearing in hydrodynamic models for describing the concentration dependence of sedimentation (Section 2.1.3) and diffusion (Section 3.3.2) coefficients are of prime importance. For these reasons, and because they arise from the same dispersion forces that cause scattering of light (see Chu, 1967, Chapter 6), they are considered in some detail.

Israelachvili (1974) has given a comprehensive review of the importance of these forces to biological systems and much of the present section comes from this work (for other discussions, see Chu, 1967, Chapter 1; and Gabler, 1978, Chapter 6). The theoretical knowledge of these forces has its basis in the work of London (1930, 1937), Hamaker (1937), Lifshitz (1956), and Dzyaloshinskii *et al.* (1961). Among other theoreticians who have expanded these views, McLachlan (1964, 1965) should be mentioned, while the careful practical work of Israelachvili and Tabor (1972) has made it possible to bridge the gap between theory and experiment.

Van der Waals' forces are less effective over large distances (greater than 5 nm) than ionic forces, and are more effective at interatomic spacings. Their energies follow a simple power law, and the relationships for specific interactions can be derived analytically. The forces are present between all molecules (even neutral molecules) and their integrated energy can exceed electrostatic interactions for charged surfaces. Since the energy is dipolar in origin (quadrupole and higher arrangements are generally small, so they can be ignored), they produce a turning moment in the atoms which tends to align two molecules until they settle in a trough formed at the minimum of the free energy of interaction. Physically, these effects can be summarized in the basic equation relating free molecular energy (G) to the distances apart (r) and polarizability (α) of the atoms in the two molecules at imaginary frequencies ($i\omega_n$) (i.e. $\alpha_1(i\omega_n)$ and $\alpha_2(i\omega_n)$). (Here $\omega_n = n(2\pi kT/\hbar)$, $\hbar$ is the modified Planck's constant, i.e. $\hbar = h/2\pi$; when $n = 0$, this term is multiplied by $\frac{1}{2}$.) Thus

$$G = -\frac{6kT}{r^6}\sum_{n=0}^{\infty}\alpha_1(i\omega_n)\alpha_2(i\omega_n) \qquad 1.33$$

The physical origin of these energies is from the effect that an external electric field (E) has on the atom or molecule. In this field, the atoms acquire a dipole moment

(μ), so $\mu = \alpha E$, where the proportionality constant, α, is the total electric polarizability of the particle. α has dimensions of volume (length^3), showing that it represents the electrical volume of the particle, while its value can be determined from the refractive index of the material (Lorentz–Lorenz equation, eq. 2.4). For mixtures, α is functionally related to the partial molar volumes, so linking it with thermodynamic functions.

Energy and polarizability

Unfortunately, α is not a simple parameter, and normally it can be factorized into three contributions. For all molecules, there exists an electronic polarizability (α_e), while, for those having permanent dipole moments μ (e.g. HCl, amino acids and aromatic molecules such as *o*- and *m*-dichlorobenzene), there are two further polarizabilities, vibrational (α_{vib}) and dipolar (α_d). In free space, $\alpha_e(\omega)$ is related to the real frequencies of oscillation of the dipole (ω) by

$$\alpha_e(\omega) = Ne^2/m(\omega_e^2 - \omega^2) \tag{1.34}$$

where e and m are the electronic charge and mass, and N is the number of effective electrons oscillating at the same average frequency, ω_e, in the ultraviolet ($\omega_e \approx 2 \times 10^{16}$ rad s^{-1}). Replacing ω by the complex frequency iω_n, gives

$$\alpha_e(i\omega_n) = Ne^2/m(\omega_e^2 + \omega_n^2) \tag{1.35a}$$

$$= \alpha_e(0)/[1 + (\omega_n/\omega_e)^2] \tag{1.35b}$$

where $\alpha_e(0)$ ($= Ne^2/m\omega_e^2$) is the static ($\omega = 0$) electronic polarizability and is essentially the polarizability in visible light ($\omega_{vis} \approx 3 \times 10^{15}$ rad s^{-1}) because $\omega_e^2 \gg \omega_{vis}^2$. For molecules having a permanent dipole moment, the polarizability of this freely rotating dipole becomes

$$\alpha_d(\omega) = \mu^2/3kT[1 - (i\omega/\omega_{rot})] \tag{1.36}$$

Again ω_{rot} is an averaged rotational relaxation frequency falling in the far infrared or microwave regions ($\omega_{rot} \approx 10^{12}$ rad s^{-1}). Replacing ω by iω_n as before, gives

$$\alpha_d(i\omega_n) = \mu^2/3kT[1 + (\omega_n/\omega_{rot})] \tag{1.37}$$

The contribution from α_{vib} is small and can generally be neglected. Thus, the relationship between total electric polarizability of a molecule in free space and frequency becomes

$$\alpha(i\omega_n) = \frac{\mu^2}{3kT[1 + (\omega_n/\omega_{rot})]} + \frac{\alpha_e(0)}{1 + (\omega_n/\omega_e)^2} \tag{1.38}$$

which at zero frequency reduces to the so-called Debye–Langevin equation, i.e.

$$\alpha(0) = (\mu^2/3kT) + \alpha_e(0) \tag{1.39}$$

These relationships (eqs. 1.35 and 1.39) can be inserted in the summation (eq. 1.33) to give (when $n=0$, $\omega_n=0$)

$$G_{n=0} = -\frac{3kT}{r^6}\alpha_1(0)\alpha_2(0) = -\frac{\mu_1^2\mu_2^2}{3kTr^6} - \frac{[\mu_{1e}(0)\mu_2^2 + \alpha_{2e}(0)\mu_1^2]}{r^6} - \frac{3kT\alpha_{1e}(0)\alpha_{2e}(0)}{r^6} \quad 1.40$$

The first term of the sum is the classical free energy of two freely rotating permanent dipoles, the second term describes the interaction between a permanent dipole and a neutral molecule (Debye energy), while the third term is the electronically induced energy (usually very small). The free energy (G) is related to the interaction energy (U) through the Gibbs–Helmholtz equation

$$U(T) = \frac{\partial(G/T)}{\partial(1/T)} \approx -\frac{2\mu_1^2\mu_2^2}{3kTr^6} - \frac{[\alpha_{1e}(0)\mu_2^2 + \alpha_{2e}(0)\mu_1^2]}{r^6} \quad 1.41$$

It can be seen from eq. 1.41 that when the temperature is low and $kT \ll (\mu_1\mu_2/r^6)$, the permanent dipoles are no longer free to rotate. Further, it shows that interaction energies between two fixed dipoles can exceed the orientation energies.

For non-zero frequencies ($n = 1, 2, 3, \ldots$), the polarizability is determined by the contributors from the electronic polarizability (α_e). These terms give the dispersion energies which, with the orientation and induction free energies, give the total van der Waals' energy and hence the forces ($F = -\partial G/\partial r$); for neutral molecules, the interactive energy arises only from dispersion forces. Although the comments given in this section must be regarded as introductory, even this would be incomplete if the point was not made that the relationship holds true only if the time taken for the field to travel between two particles is small in comparison to the time it takes for one electron to rotate about the proton ($c/r > \omega_c$ where c is the velocity of light in vacuum). As c/r approaches and eventually exceeds ω_c, then the frequencies greater than c/r are damped and no longer contribute to the dispersion energies. These are called retardation effects, and the van der Waals' energy is decreased.

Energies of molecules in solution

These energies would only have theoretical interest to general discussions on the behaviour of macromolecules if it were not for the fact that from them arise hydrophobic effects between neutral solvent and solute molecules. They are also responsible for the production of stable arrays of large particles in solution, where electrostatic repulsions are balanced by the attractive van der Waals' forces.

Discussing the hydrophobic effect (see Section 1.1) from this aspect requires that eq. 1.33 be modified to include a dielectric permeability, $\varepsilon_3(\mathrm{i}\omega)$, for the solvent (again at imaginary frequencies). This new term reduces the forces between the atoms, so making the polarizabilities different from free-space polarizabilities. Further, the retardation effects become important at shorter distances; this follows because the speed of light is reduced by $\varepsilon_3^{1/2}$.

From polarizability could be calculated in principle virial coefficients for neutral molecules, but in practice it becomes difficult to ascribe values to the excess polarizabilities of a molecule (α^*) in solution. The value of α^* depends in a complicated way on the type of interaction between solvent and solute. McLachlan (1965) has described a relationship for the excess polarizability of a spherical molecule in a continuous medium. This shows that the forces between the particles are much reduced in the presence of a solvent, and for small molecules the energies are small compared with kT. Increasing the size of the particle considerably increases this energy, but again detailed analyses can only be made for models of known and simple geometrical shape. For example, Israelachvili (1974) quotes the case of a small anisotropic particle interacting with an interface having dimensions which can be regarded as large relative to those of the small particle, when the energy is given by

$$G = -\frac{kT}{2r^3} \sum_{n=0}^{\infty} \alpha_1^*(\mathrm{i}\omega_n) f(\varepsilon_3(\mathrm{i}\omega_n)) \qquad 1.42$$

where $f(\varepsilon_3(\mathrm{i}\omega_n))$ is a function of relative dielectric permeabilities.

Comparing eq. 1.42 with eq. 1.33 shows that the energy decreases more slowly with distance for interfacial interaction than with small point charges ($G \propto r^{-6}$), and under some circumstances the relationship approaches that found for electrostatic energies ($G \propto r^{-1}$) as two interfaces are brought near to one another ($G \propto r^{-2}$ or even $G \propto r^{-1}$). These cases are of considerable interest in biology. Possibly the best-known experimental phenomenon explained through balancing electrostatic repulsion against van der Waals' attractions is the one found when tobacco mosaic virus (TMV) is dissolved in dilute salt solutions. These rod-like molecules (approximately 300 nm by 18 nm) form stable arrays at concentrations of about 18 g l^{-1} in dilute salt (0.01 M to 0.2 M). The distance between the rods varies both with ionic strength and with pH, but for any fixed condition the spacings are stable. Using X-ray crystallography, Bernal and Fankuchen (1941) measured these spacings, showing that the virus arranged itself in hexagonal assembly with centre-to-centre distances varying from 17 nm to 100 nm depending on conditions. Many attempts have been made to formulate models which describe this balance between electrostatic and van der Waals' energies to produce a resulting minimum in the energy–distance relationship. One general form of this type of curve for a cylindrical macromolecule is given in Figure 1.5 (taken from Brenner and McQuarrie, 1973), showing that the molecules would arrange themselves in an array having between 7 and 10 nm distance between individual rods. The distance increases with increasing pH. Earlier workers studying the stable stacking of TMV utilized the classic theory of

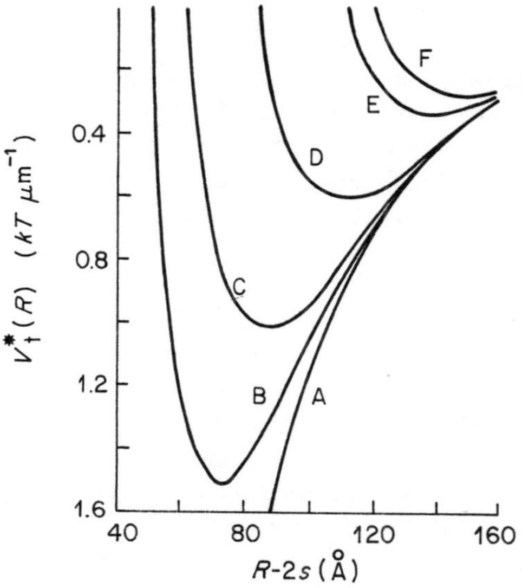

Figure 1.5 The relationship between total reduced attractive energy ($V_t^*(R)$) and distance between cylindrical particles ($R - 2s$, where s is the radius of the cylinder). Values are given for several pH values at ionic strength 0.05 M. Curves A–E are for pH 4.0, 4.4, 4.6, 5.0 and 6.0, respectively, while curve F is for pH 7, 8, 9 and 10. (Taken from Brenner and McQuarrie, 1973, *Biophys. J.* **13**, 301, Figure 10. Reproduced by permission of the Biophysical Society)

Derjaguin, London, Verwey, and Overbeek (D.L.V.O.) to describe the formation of the stable array, but, as with modern interpretations, the models depend on the principle of a balance between attractive and repulsive forces. Although these calculations qualitatively describe the interactions, attempts to quantify the work have been frustrated through insufficiently accurate experimental data, as well as through difficulties in ascribing accurate values to either polarizabilities of the molecules or correct functions of $\varepsilon_3(i\omega)$ (Parsegian and Brenner, 1976). Other biological systems that produce ordered arrays in nature which can be modelled on these interactions are the A-band lattice of striated muscle (Brenner and McQuarrie, 1973), the visual cornea, and the gel of sickle cell haemoglobin (Minton, 1973).

More space has been devoted to van der Waals' forces than to the other energies because of their ubiquitous distribution. Later, the dispersion formulae will be seen to apply to light scattering (Chapters 2 and 3). In general, these energies are the source of many interactions between macromolecules and, although quantitative descriptions of the interactions may still remain elusive, it is generally accepted that a balance between the energies of dipole interactions and electrostatic forces can produce stable aggregates in solution.

1.5 Conclusions

The topics covered in this chapter were chosen to give a canvas upon which details of macromolecular behaviour can be drawn. For this reason, the general treatments usually employed to describe properties of both large and small molecules have been outlined, rather than taking any one topic and treating it in detail. This superficial discussion shows that detailed knowledge of the interactions between molecules occurs at several levels of complexity depending on the size of the molecule and on whether or not solvent is present. For macromolecules, it is difficult to extend theory beyond qualitative ideas, but even these are useful when trying to interpret experimental results. In order to emphasize these influences, it is worth listing the four basic facts that emerge from the discussions of forces between molecules and which form the underlying principles for viewing subsequent discussions.

(1) In aqueous solutions, the weak bonding of water to macromolecules provides a stabilizing environment whose energy determines the solubility of part or all of the macromolecule. The inevitable outcome from these interactions is that the solution must be considered as a whole unit, and to discuss only the one component of the mixture, while ignoring its influence on the solvent, cannot provide a true account of the behaviour of the macromolecule.

(2) It follows logically from (1) that, if exact formulations of solvent–solute interactions are not possible, the interpretations of experimental facts must include semi-empirical coefficients. Generally, these coefficients are required for adjusting thermodynamic relationships and are called virial or activity coefficients, but their common origin lies in interactions deriving from ionic, hydrogen-bond, and van der Waals' energies.

(3) Of the bonding energies available for forming macromolecules, the covalent bond provides the strongest force between atoms. The result is a polymer consisting of a stable matrix of varying degrees of stiffness. It may seem strange, therefore, to find that biologically it is the secondary and weaker forces that determine whether a synthesized macromolecule is finally active. In many circumstances, ionic interactions are the strongest of the secondary energies, followed by hydrogen bonding and van der Waals' forces. Balanced attractive and repulsive forces between these three sources of energy determine the extent of solvation, tertiary configuration, and flexibility of the macromolecule. Despite knowing the exact physical origins of these forces, it is impossible to give a general statement on the relative energies of their interactions for macromolecules: this varies with size and solvent.

(4) Since the solvent and solute form a complete system, no single experimental treatment can discriminate between the forces. Further, it is unlikely that absolute proportioning of the total energy of the system between the various weak forces is of real importance to the understanding of biological activities of macromolecules. Usually, it is the differences in these proportions caused by a chemical reaction that are important. These changes can be described as flexing, aggregation, or change in volume of the macromolecule.

CHAPTER 2

Weighing molecules

Historical trends

The weight (or relative molecular mass) of a macromolecule was an early noteworthy physical parameter to be accurately measured in biochemistry—oxyhaemoglobin was the molecule measured by Svedberg and Fåhraeus (1926). Following this historic experiment, the relative masses of many proteins were measured in the hope that a fundamental unit existed from which all proteins were assembled. This unit was thought to vary in composition but possess a mean relative mass of about 17 000 dalton. Hindsight shows that it would have been surprising if all proteins could be assembled from one standard unit. It is interesting to note, however, that polymer chemists are well aware of the almost impossible task of synthesizing linear chains guaranteed to have no cross-links and containing more than about 100 monomeric units. Side reactions, which normally are insignificant when related to a linear polymerization, become important as the number of monomers in the polymer increases (the probability of forming a cross-link increases with the time of exposure): one cross-link in 100 linear linkages gives a branched polymer instead of the expected linear chain. This statistical problem is probably responsible for the fact that amylose, the so-called linear starch molecule, is found to have occasional branches. Thus, it could be argued that constructing large macromolecules by precise cross-linking of small chains is the only way to build routinely large molecules of reproducible structure. These chains could be cross-linked through covalent bonds such as disulphide bridges. Viewed in this light, the ideas of early protein chemists were not so far from reality, although they would have had to break disulphide groups in order to test the hypothesis and this was not tried. Nevertheless, the accumulation of data on relative masses of a wide variety of proteins laid the foundation of biophysical chemistry and the results showed that the solvent–solute interactions given for small molecules could be extended to proteins in solution.

Nucleic acids proved to be even more intractable than proteins in their fractionation, and the recognition that they were composed of well defined entities occurred much later than for proteins. The early studies of nucleic acids were largely confined to the organic chemistry of the nucleotides, and it was not until the late 1940s that methods were found which separated mixtures of nucleic acids into discrete species. It was then found that both deoxyribonucleic and ribonucleic acids could be isolated as well defined macromolecules each possessing unique biological properties. Their separate classifications were by their sedimentation rates, because there was no simple procedure available for testing their biological activities. Unlike the proteins, there was never any suggestion that nucleic acids could be assembled from a standard small linear chain; it was recognized early that, in order for nucleic acids to preserve their coded information, the synthesis must occur in one chain. Presumably, the biological trick used to maintain linearity here is to synthesize the new chain on a template (the old chain) and so considerably reduce the chances of random cross-linking through low-level side reactions.

Fragility of macromolecules

The early work on fractionating nucleic acids led to many controversies concerning their real size; it was not realized that the covalent bond linking the nucleotides was relatively fragile. Simple laboratory procedures, such as stirring a solution, broke at least one bond in the linear chain, and, since it is more probable to break a bond in the middle than at the ends, the effect is to halve the size of the molecule. This process contrasts markedly with proteins, which are relatively stable under such mild treatments (although vigorous mixing will denature proteins as a result of producing a large surface excess energy in the increased surface areas of the bubbles, and this energy can unfold globular proteins), and may be ascribed to the fact that nucleic acids adopt a conformation in solution which approximates to a random coil. Not only does this conformation mean that most, if not all, of the units in the chain are exposed to the solvent but also, since they are very large molecules, keeping one end stationary, through interaction with a stationary region of solvent, will introduce considerable strains if the other end moves: this strain is relieved by hydrolysis. This simple description is the one normally given to explain the fragility of nucleic acids and, although it is obviously important, there is a further equally important factor, namely the requirement that nucleic acids are often studied in alkaline solutions. This was particularly prevalent in the early days of the studies. When alkaline solutions are oxygenated, they contain relatively high concentrations of free radicals ($OH^{\cdot}$ and $O^{\cdot}$) and these will react with any bonds. A very effective experimental demonstration of the destructive effect that these radicals have on molecules bearing ether links is to bubble air through a hot alkaline solution of starch and to measure its viscosity at various times (Gilbert, 1958). The viscosity of both amylose and amylopectin decrease dramatically in this experiment, showing that the molecules are being chopped into smaller pieces. Yet substituting pure nitrogen for the air and repeating the experiment gave no corresponding drop in relative masses. Thus, for nucleic acids, it seems reasonable to suppose that the effect of mechanical strain in the presence of large concentrations of free radicals led to hydrolysis of many bonds.

Among the many other groups of biological macromolecules which exist widely in nature are lipoproteins, glycoproteins, and polysaccharides, but these remain polydisperse macromolecules which have defied efforts of subclassification through principles similar to those used in the study of proteins and nucleic acids. It seems that the essence of these polymers is polydispersity. It could be that our views about these polydisperse macromolecules will appear crude in future assessments, but the attempts made to fractionate them into separate units having unique biological activities have been extensive and largely unsuccessful, so details about these macromolecules are largely omitted from this discussion. The exclusion of polysaccharides and lipoproteins reflects the view that, since they occur largely as food reserves or physical supports of uncertain biological role, it is difficult to assay their biological activity in order to relate physical parameters to function. Apart from gross divisions based on the presence or absence of cross-links as with amylose and amylopectin for starch, polysaccharides are generally considered to be natural polymers produced by condensation of simple monomers. Their size is determined mainly by the relative kinetics of the reactions rather than by predefined criteria for activity. Lipoproteins and glycoproteins may have important biological roles but, from the standpoint of physical chemistry, they are complex mixtures which are often difficult to separate into pure macromolecules.

Although accurately determined relative molecular masses formed the keystone in the development of physical biochemistry, recent developments have reduced the need for accurate physical measurement of mass. It can be argued that once a pure fraction has been obtained then sequencing will immediately give the accurate relative molecular mass. It is true that an approximate molecular mass aids sequencing, but this can be obtained without resorting to elaborate physical methods. These methods depend on a detailed knowledge of physical theory and maintenance of expensive equipment: these factors reduce the incentive to explore accuracy. The wealth of detail required to interpret results from a physical measurement must in itself thwart researchers who only wish to decide whether a molecule is of relative mass 17 000 dalton or 170 000 dalton. Simple divisions of this sort can be readily made by gel electrophoresis, but they can never provide data on reactions between macromolecules which lead to absolute determination of their mutual interaction coefficients. Normally, the amino acid sequence will not provide a relative molecular mass of an active molecule but only that of the smallest subunit, and the results can never be used to measure the strength of binding between the subunits (although differential gel chromatography can yield accurate measurements of differences in binding energies, and cross-linking chains with bifunctional reagents indicate the number of chains in an oligomer). Similarly, these relative methods can never be applied to studies of the concentration dependence of parameters and, since these relationships are the only source of measuring the weak interactions between solutes themselves as well as with the solvent, this means that there will always be a place for accurate determinations of relative molecular masses: they complement the other hydrodynamic measurements such as diffusion (Chapter 3).

Generally, the properties of biological macromolecules are studied after the solid has been dissolved in a buffered solvent containing small quantities of ions, macromolecules, and a vast excess of water (in subsequent equations, they are referred to by subscripts 3, 2, and 1, respectively). The very low vapour pressure of natural polymers excludes the use of conventional mass spectrometers for determining their relative molecular masses. Instead, work is confined to measuring suitable properties of solutions which can yield relative molecular masses; these are not dependent on the specific properties of the solute but only on the change in chemical activity of the water induced by dissolving a known but small quantity of the solute (see Chapter 1). This effect is best described by the second law of thermodynamics which led to the Gibbs–Duhem relationship (eq. 1.5). Thus, the mole fraction of the solute is the important experimental parameter; from this can be calculated the relative molecular masses once the weight of solute per unit weight of solvent is known. The change in solvent activity is expressed macroscopically in vapour-pressure depression, elevation of boiling point, depression of freezing point, and osmotic pressure of solutions relative to the pure solvent, but in the centrifuge it is the solute activity which is measured. Obviously, boiling-point elevation is useless for biological macromolecules because of thermally induced interactions between the macromolecules. Using the molar freezing-point depression for water (1.86°C mol^{-1}), calculation shows that the depression of freezing point of a 1% w/v aqueous solution containing a macromolecule of relative mass 20 000 dalton is about 10^{-2}°C, decreasing as the relative mass increases. Hence, it is obvious that freezing-point measurements are not going to give accurate measurements of relative mass for macromolecules. Similar calculations for vapour pressure show that this macromolecular solution would have a depression of about 10^{-3} mm Hg; low, but possible to measure by differential manometry. Finally, the same solution would produce an osmotic pressure of between 10 and 20 cm H_2O. This method can therefore be applied successfully to solutions of macromolecules (see Chapter 1). Its disadvantage, if accuracy is to be maintained, is that the operational concentration must be increased with increasing relative mass, making it unsuitable for studying large macromolecules.

Certain physical methods are able to break away from the limits set through dependence on colligative properties of solutions. Svedberg, for example, constructed a novel mass spectrometer which depended on centrifugal forces to separate and weigh molecules rather than using the differing charges and masses as with a conventional mass spectrometer. Others have developed the measurement of scattered light to determine mass–density in solutions. Although these methods appear to depend on physical, rather than chemical, principles, it is important to remember that the activity of the solute and solvent are still the measured parameters and that the relative molecular mass is calculated from them.

2.1 Survey of methods for estimating relative molecular masses

Methods fall into two complementary divisions, those that provide absolute parameters and those that give relative values. The first group includes osmotic pressure (Chapter 1), light scattering (also see Chapter 3), and ultracentrifugal techniques; the relative methods include gel chromatography and gel electrophoresis. It is only possible to measure relative molecular masses using the gel procedures if absolute masses of macromolecules are available to use as standards.

2.1.1 Osmotic pressure

The basic theory of osmotic pressure is given in Chapter 1, but, since no discussion is included of the specific applications to macromolecules, this method is considered further here. Since this is one of the older physical methods used to measure relative molecular masses of macromolecules, a large literature exists describing results, theory, and practice; fortunately, these have been summarized by Tombs and Peacocke (1975), so it only remains for this section to introduce the use of osmotic pressure as a method for measuring masses and virial coefficients of macromolecules. Generally, osmotic pressure measurements made on macromolecular solutions are not as accurate as those made with the ultracentrifuge (see Section 2.1.3) when concentrations are less than $1\,g\,l^{-1}$, but valuable contributions have been made using osmotic pressures of higher concentrations. At low concentrations, the hydrostatic corrections introduce uncertainties which make absolute measurements unreliable. An order-of-magnitude calculation of the osmotic pressure generated by a 0.1% w/v solution of a macromolecule of relative molecular mass 66 000 dalton at 20°C gives a value of about 3.8 mm H_2O. This is measured as a rise up a capillary tube, and the recorded experimental rise will be greater than this and must be corrected for surface tension. For pure water, this correction is calculated from the formula (0.15 ÷ radius of capillary) and in the present example the correction equals the expected rise caused by osmotic pressure when the radius is about 4 mm, which is a large bore. Using a 1 mm bore, the total rise is about 4.7 mm. Thus, a 10% error in estimating the correction gives a final error of about 3% in the corrected relative molecular mass (to give 64 000 dalton). Hence, for this size of molecule and these concentrations, osmotic pressure gives an acceptable accuracy. Lowering the concentration below 0.1%, or raising the relative molecular mass, decreases the accuracy in proportion because the magnitude of the correction remains relatively constant.

Non-ideality of haemoglobin solutions

Raising the concentration of the solute increases the deviations of the solution from thermodynamic ideality, but this is true of all methods of measuring molecular masses. This in itself is not undesirable since many biological macromolecules exist physiologically in high concentrations, but it does mean that it is necessary to understand the origin of the deviations in order to correct results. Only then can realistic relationships between biological activity and physical activities be discussed. Haemoglobin in erythrocytes is possibly a good example, where the concentration of a globular protein is normally greater than

300 g l^{-1}: if organs such as muscles were included as a protein solution, then concentrations greater than this would be biologically normal. At a concentration greater than 300 g l^{-1} (w/v), the unsolvated haemoglobin molecule would occupy more than a quarter of the volume of solution, and if solvation is included then the remaining free volume of water would be very small. This means that the chemical activity of the solvent is considerably reduced. Under these conditions, measurements of osmotic pressure provide accurate data for assessing solute–solvent interactions covering a wide range of concentrations. As early as 1928, Adair (1928) measured the osmotic pressure of sheep haemoglobin in concentrations extending up to 350 g l^{-1}. Attempts to relate the deviations from ideality to molecular size and activity coefficients have had moderate success, but possibly the most interesting attempt was made by Ross and Minton (1977). Equation 1.16 (the relationship between osmotic pressure and concentration, which includes virial coefficients) can be rewritten in the form

$$\Pi = \frac{RT}{M_2}\left(w_2 + \sum_{i=2}^{n} \Gamma_i w_2^i\right) \tag{2.1}$$

and using eq. 2.1 each Γ_i can be estimated from the experimental record by statistical regression analysis, providing the coefficients are treated as uncorrelated parameters. This method may be statistically sound but it does not provide a physical interpretation of the parameters. Hence, Ross and Minton (1977) rephrased the problem by presuming that the long-range electrostatic interactions between separate tetrameric haemoglobin molecules could be ignored at all concentrations. Then non-ideality could be equated with the mutual impenetrability of compact solvated haemoglobin molecules, that is, the mutual exclusion principle discussed in Chapter 1. On this basis, they found that Adair's results could be adequately represented by a polynomial containing seven terms (Figure 2.1), but these coefficients were not all estimated by empirical fitting. Instead, the higher-order terms ($i = 3, 4, \ldots$, eq. 2.1) were calculated from Γ_2 using the methods of Ree and Hoover (1967) and Hirschfelder *et al.* (1967), leaving M_2 and Γ_2 as the only coefficients to be estimated from the data. As can be seen from Figure 2.1, the fitting was good, although the value estimated for M_2 (68 000 dalton) was possibly higher than expected for tetrameric haemoglobin but in agreement with Adair's original estimate. The older estimate of relative mass was made independently using measurements at low concentrations where the solution was essentially ideal in behaviour (the relative molecular mass of sheep haemoglobin calculated from the amino acids plus haem is 64 442 dalton).

The model used by Ree and Hoover (1967) to calculate high-order coefficients from Γ_2 assumes the value of Γ_2 is proportional to the specific volume of the molecule. Ross and Minton (1977) show that if haemoglobin is represented as a cube having a face diagonal of 6.2 nm, then the calculated value of Γ_2 is similar to the measured second virial coefficient. Arguments that the model is biased by not including long-range interactions is countered by showing that, if there are 15 charges per molecule, the electrostatic energy contributes only 15% of kT, and an insignificant factor in the non-ideality of oxyhaemoglobin solutions. In fact, this

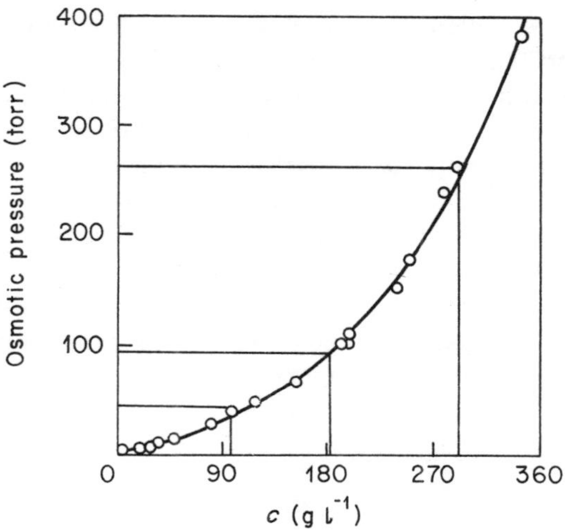

Figure 2.1 Relationship between osmotic pressure and concentration of sheep haemoglobin, plus KCl, ionic strength 0.3 M, pH 7.8 at 0°C: ○, data from Adair (1928); the continuous line is the curve fitted according to eq. 2.1 from coefficients calculated using $B_2 = 3.83 \times 10^{-3}\,\mathrm{l\,g^{-1}}$ and relative mass of 68 000 dalton. (Taken from Ross and Minton, 1977, Figure 1. Reproduced with permission from *J. Molec. Biol.* **112**, 437. Copyright by Academic Press Inc. (London) Ltd)

conclusion must also apply to many globular proteins at pH 7, where the number of residual charges seldom exceeds 20. Finally, this work makes it possible to calculate the numerical relationship between chemical activity and concentration. At 400 $\mathrm{g\,l^{-1}}$, the thermodynamic activity is over two orders of magnitude greater than its weight concentration (conversely, the activity of water is two orders of magnitude less than its concentration). Thus, the biological activity of haemoglobin will be enhanced by these orders of magnitude compared with that expected by calculating simple weight concentrations (further discussions of solute–solvent interactions for haemoglobin are given in Chapters 3 and 4).

2.1.2 Intensity light scattering

Scattering of electromagnetic radiation from atoms and molecules is an everyday phenomenon, producing the red sky at night and the blue of the seas (these effects are called Rayleigh scattering after Lord Rayleigh who, in 1881, described the scattering mathematically). Physically, light-scattering theory starts by discussing the interactions between electromagnetic waves and oscillatory motion of electrons in the constituent atoms of the scatterer. The intensity of the scattered light depends on the mass of atoms per unit volume, and from this can be calculated a relative molecular mass. Extensive applications of this concept for measuring molecular parameters have been made, but with the advent of lasers its scope has widened. Results before about 1970 were from measurements of the intensity of the scattered light (elastic scattering), whereas after this date studies were made more frequently using interferometric methods to measure shifts in frequency of the scattered light caused by random motion of the particles (quasi-elastic light scattering, Chapter 3). Molecular masses can be calculated directly from the intensity method, and quasi-elastic light scattering measures Brownian motion which is only indirectly related to mass through a shape factor. Hence, it is simpler to consider elastic or intensity scattering here and to consider quasi-elastic scattering when discussing shape (Chapter 3).

Elastic scattering

Tanford (1961, Chapter 5) gives a concise account of the theory of elastic light scattering; a more comprehensive account covering all aspects of scattering is given by Chu (1974). These discussions start with the polarizability of a molecule (α, see Chapter 1). This parameter represents the oscillatory electrical volume of the molecule and provides a measure of the cross-sectional area of the molecule through which interaction with the external field occurs, i.e.

$$\mu = \alpha E \qquad 2.2$$

The dipole moment of the particle (μ) is related to the incident field strength (E) through eq. 2.2. If the incident beam is polarized, then the basic Rayleigh scattering equation for an isolated gas molecule is

$$\frac{I_s}{I_0} = \frac{16\pi^4\alpha^2\sin^2\theta}{\lambda^4 d^2} \qquad 2.3$$

where the wavelength λ (*in vacuo*) is that of the incident beam of intensity I_0. The intensity of the scattered light (I_s) is measured at angle θ at a distance d from the isotropic gas molecule.

Thus, from light-scattering experiments is given an estimate of α, and this can be converted into molecular mass through a further relationship between α and molecular mass. Normally, the one used is the Lorentz–Lorenz relationship which relates α to the refractive index ($\bar{n}$) given by

$$\bar{n}^2 - 1 = 4\pi N\alpha \qquad 2.4$$

where N is the number of particles per unit volume. For pure gases at low pressures, $\bar{n} = 1$. Hence, if $\bar{n}$ is measured at different partial pressures for a mixture of gases (which, for solutions, are equivalent to differing concentrations), then its relationship with concentration can be formed by expanding ($\bar{n}^2 - 1$) as a Taylor series in which high-order terms are neglected to give

$$\bar{n} = 1 + \frac{\mathrm{d}\bar{n}}{\mathrm{d}c}c \qquad 2.5$$

where c, the concentration (weight per unit volume) is related to N, through Avogadro's number ($\mathcal{N}$) and the molecular weight of the particle. Thus, substituting for n from eq. 2.4 into eq. 2.5 and combining with eq. 2.3 gives

$$\frac{I_s}{I_0} = \frac{4\pi^2 \sin^2 \theta (\mathrm{d}\bar{n}/\mathrm{d}c)^2 M_2 c}{\mathcal{N} \lambda^4 d^2} \qquad 2.6$$

Equation 2.6 provides the basic equation for relating scattered polarized light with molecular mass. If the incident beam is unpolarized, then I_0 contains two terms, one for each component of the light, and eq. 2.6 becomes

$$\frac{I_s}{I_0} = \frac{2\pi^2 (1 + \cos^2 \theta)(\mathrm{d}\bar{n}/\mathrm{d}c)^2 M_2 c}{\mathcal{N} \lambda^4 d^2} \qquad 2.7$$

Equation 2.7 only applies to scattering from simple gaseous mixtures at low pressures (ideal conditions). With scattering from a binary mixture, both components scatter light, making it necessary to replace the '1' in eq. 2.5 by the refractive index of the solvent ($\bar{n}_S$) to give

$$\bar{n}_S^2 - \bar{n}_0^2 = 2\bar{n}_0 (\mathrm{d}\bar{n}/\mathrm{d}c) c \qquad 2.8$$

which when used to replace α in eq. 2.3 gives

$$\frac{I_s}{I_0} = \frac{2\pi^2 (1 + \cos^2 \theta) \bar{n}_0 (\mathrm{d}\bar{n}/\mathrm{d}c)^2 M_2 c}{\mathcal{N} \lambda^4 d^2} \qquad 2.9$$

Equation 2.9 is the practical equation used to study light scattering from a thermodynamically ideal solution and contains six variables which have to be measured (I_s/I_0, θ, $\bar{n}_0$, $\mathrm{d}\bar{n}/\mathrm{d}c$, c and d, assuming λ is known) before M_2 can be calculated. Of these six, $\mathrm{d}\bar{n}/\mathrm{d}c$ must be measured as accurately as possible and under identical conditions to those empoyed in the experiment (Eisenberg, 1976, p. 59). The error in M_2 is proportional to $(\mathrm{d}\bar{n}/\mathrm{d}c)^{-3}$ and, since $\mathrm{d}\bar{n}/\mathrm{d}c$ is less than unity, the proportional error increases with decreasing $\mathrm{d}\bar{n}/\mathrm{d}c$. However, the intensity of the scattered light increases with $(\mathrm{d}\bar{n}/\mathrm{d}c)^2$ and, since $\mathrm{d}\bar{n}/\mathrm{d}c$ will be larger for an expanded molecule than for its compact spherical equivalent (where solvent is excluded from inside the sphere), the accuracy of light scattering improves with extension of the polymer chain. Absolute values of $\mathrm{d}\bar{n}/\mathrm{d}c$ will be affected by the extent of interactions between diffusible counter-ions and the macromolecule; the effects are similar to those occurring when partial specific volumes are measured (see Section 2.1.3). For globular particles having dimensions less than about 0.05λ, the dependency of I_s/I_0 on θ is not important (for visible light, this limits it to a maximum ordinate of about 23 nm). When the dimensions exceed this value, interference between scattered waves produced by spatially separated oscillators reduces I_s/I_0 at all angles except when $\theta = 0$. This property is useful when determining average dimensions for large particles (see p. 87).

Non-ideality in light scattering

Thermodynamically ideal solutions are never realized in practice and, as with osmotic pressure, corrections for non-ideality must be included when working with finite concentrations. Non-ideality introduces cross terms between the polarizabilities ($\alpha_{1,2}$) which must be included as well as $\alpha_{1,1}$ and $\alpha_{2,2}$ in eq. 2.3 (eq. 2.8 assumes only $\alpha_{1,1}$ and $\alpha_{2,2}$apply). Equation 2.9 can be expanded (Tanford, 1961, p. 285) by including the influence that local fluctuations in concentration in the illuminated volume element have on the scattering, to give

$$\frac{I_s}{I_0}=\frac{2\pi^2\bar{n}_0^2(\mathrm{d}\bar{n}/\mathrm{d}c)^2(1+\cos^2\theta)c}{\lambda^4 d^2[-(1/\bar{V}_1 kT)(\partial\mu_1/\partial c)_{T,P}]} \qquad 2.10$$

Here V_1 is the partial molar volume of the solvent of chemical potential μ_1. The terms in the square bracket in the denominator of eq. 2.10 are equivalent to the virial coefficients of the gas law (eq. 1.20) and can be replaced by

$$\mathcal{N}[(1/M_2)+2B_2c+3B_3c^2+\cdots]$$

(eq. 1.24) where M_2 is the true relative molecular mass of the solute. The virial coefficients (B_i) correspond in every way to those from osmotic pressures (eq. 1.16). Thus, the complete light-scattering equation becomes

$$\frac{2\pi^2\bar{n}_0^2(\partial\bar{n}/\partial c)_{T,P}}{\mathcal{N}\lambda^4}\,\frac{I_0(1+\cos^2\theta)}{I_s d^2}=\frac{1}{M_2}+2B_2c+\cdots \qquad 2.11$$

The experimental procedures used to measure light scattering have been described by several authors and need not be repeated in detail here (Pitz *et al.*, 1973; Evans, 1972; Eisenberg, 1971). When studying small particles, it is normal to measure the scattered intensity at either a fixed angle (90° relative to the incident bean) or range of angles over a range of macromolecular concentrations. Then, assuming $\mathrm{d}\bar{n}/\mathrm{d}c$ is constant, the first part of eq. 2.11 is an optical constant (Ω), the second part is experimentally determined (R_θ), and the third is the concentration. Thus, plotting $\Omega c/R_\theta$ against c should produce a straight line whose slope is equal to $2B_2$ and whose intercept equals $1/M_2$. If the particulate dimensions approach those of the wavelength of light (which for visible light is around 500 nm), then more complex analyses are required (Tanford, 1961, p. 297). From these analyses, it is possible to calculate the molecular mass, virial coefficients, and average radius of gyration of the particle (Zimm plot; Zimm, 1948; see Section 2.5).

When the solution contains a mixture of macromolecules, it can be shown that light scattering yields a weight-average relative molecular mass ($\bar{M}_w$) at infinite dilution where

$$R_\theta=\Omega\sum_i c_iM_i \qquad 2.12$$

(c_i is the concentration of the ith component, having a relative molecular mass M_i, assuming $\mathrm{d}\bar{n}/\mathrm{d}c$ is the same for all species). From the plot of $\Omega c/R_\theta$ against c (the total concentration equals $\Sigma_i c_i$), we have

$$\frac{\Omega c}{R_\theta} = \frac{\Sigma_i c_i}{\Sigma_i c_i M_i} = \frac{1}{\bar{M}_\mathrm{w}} \tag{2.13}$$

The limiting slope of the plot gives a Z-average virial coefficient and not a weight-average value (this is obtained from osmotic pressure measurements of polydisperse solutions).

2.1.3 Ultracentrifuge

Following Svedberg's demonstration that the ultracentrifuge could resolve a mixture of macromolecules into separate components, as well as provide estimates of their relative molecular masses, the instrument became common property in biochemical laboratories. Modern preparative centrifuges can produce accelerations exceeding a half-million times that of the Earth's gravity, while analytical models approach this. The two types of centrifuge are similar in construction, but analytical centrifuges include optical trains for recording the positions of the sedimenting boundaries as well as accurate monitoring systems for speed and temperature. The descriptions given by Schachman (1959), Creeth and Pain (1967), and Lloyd (1974) cover the modern practical and basic theory of analytical ultracentrifugation. These texts have superseded the original monograph by Svedberg and Pederson (1940), but this still remains a primary source of early work. In addition to these reviews, there exist collections of research papers describing the application of the ultracentrifuge (Williams, 1963; *Annals of the New York Academy of Sciences*, **169**; *Biophysical Chemistry*, **5**; *Methods in Enzymology*, **27** D).

The analytical ultracentrifuge can be operated in two contrasting experimental modes; the velocity mode, where the velocity and shape of a sedimenting boundary are studied, and the equilibrium mode, which employs slower speeds of the rotor to produce equilibrium between diffusional and centrifugal potentials. The velocity mode separates components in a mixture, but does not give relative molecular masses directly. These are more readily obtained from the equilibrium mode. This latter mode is not useful for separating components and is relatively less reliable for detecting heterogeneity. Theories of centrifugation include two molecular parameters in addition to the relative molecular mass, namely the sedimentation constant ($s^\circ_{20,\mathrm{w}}$) (defined as the velocity of the solute per unit acceleration at infinite dilution in water at 20°C) and the diffusion constant ($D^\circ_{20,\mathrm{w}}$) (defined as the mass flowing through a unit plane by random motion of the molecules, again normalized to water at 20°C; also see Chapter 3). The sedimentation constant $S^\circ_{20,\mathrm{w}}$ is a transport constant having a basis in hydrodynamics and irreversible thermodynamics but not in equilibrium thermodynamics, while diffusion derives its potential from the relative chemical activity of the solute compared with the solvent and for equilibrium centrifugation is a thermodynamic parameter. Diffusion and sedimentation are related through the drag or frictional coefficient (see Section 3.2), and this coefficient is also related to mobility in electrophoresis. Thus, all these transport properties are linked through the shape of the molecules. The fundamental equation of sedimentation is the differential equation named after its formulator, Lamm (1929), i.e.

$$\left(\frac{\partial c}{\partial t}\right)_r = D^\circ\left(\frac{\partial^2 c}{\partial r^2} + \frac{1}{r}\frac{\partial c}{\partial r}\right) - \omega^2 S^\circ\left(r\frac{\partial c}{\partial r} + 2c\right) \qquad 2.14$$

where ω^2 is the squared angular velocity ($2\pi \times \text{rev sec}^{-1}$) and t is the time. Lamm's equation describes the temporal distribution of concentration with radius for a sedimenting molecule in a binary mixture (solvent and solute). From eq. 2.14 can be derived the working formulae for sedimentary velocity and equilibrium (see Section 2.7.1), namely

$$M_2 = \frac{RT(\partial c/\partial r)_r}{(1 - \bar{V}_2\rho_S)\omega^2 rc} \qquad 2.15a$$

$$M_2 = \frac{2RT\,\mathrm{d}\ln c(r)}{(1 - \bar{V}_2\rho_S)\omega^2\mathrm{d}(r^2)} \qquad 2.15b$$

$$M_2 = \frac{RT \quad S^\circ}{(1 - \bar{V}_2\rho_S)\,D^\circ} \qquad 2.16$$

where $\bar{V}_2$ is the partial specific volume and ρ_S is the density of the solution. Equation 2.16 can also be derived from simple transport considerations (Svedberg, 1925).

It is normal not to include subscripts when writing eqs. 2.15 and 2.16 but it should be remembered that solutions of biological macromolecules often contain relatively high concentrations of buffer ions which also undergo sedimentation. Thus, a complete formulation of the Lamm equation should contain terms which allow for the sedimentation of these ions. But, at the practical accelerations possible, only negligible sedimentation of these small molecules occurs and thus their inclusion can be neglected. The redistribution of small solutes at sedimentation equilibrium is only important when they contribute to the density changes down the cell.

Velocity ultracentrifugation

The sedimentation coefficient can be estimated from records by determining the velocity of the centre of mass of the boundary ($\bar{r}$) (Goldberg, 1953), when S can be calculated from

$$\frac{\mathrm{d}\ln\bar{r}}{\mathrm{d}t} \equiv \omega^2 S \qquad 2.17$$

Correcting S for viscosity of the solvent and density of the solution relative to water at 20°C (η° and ρ°, eq. 2.18) gives $S_{20,\mathrm{w}}$ as

$$S_{20,\mathrm{w}} = S\frac{(1 - \bar{V}_{20}\rho^\circ_{20})\ \eta_T}{(1 - \bar{V}_T\rho_T)\ \eta_{20}} \qquad 2.18$$

Equation 2.18 can be used when the partial specific volumes are not significantly altered by changing temperature and concentration. The volume is affected by the presence of other small molecules in solution, and inaccurate allowance for these interactions can produce large systematic errors in both $S_{20,w}$ and relative molecular mass. A 1% error in $\bar{V}$ produces a 2.7% error in the ratio (eq. 2.18). In order to appreciate the problems arising when attempting to measure $\bar{V}$ at this level of accuracy, it is necessary to discuss the origin of $\bar{V}$ as a partial molar quantity (see Chapter 1). From there, it can be defined as

$$(\partial\rho/\partial c_2)_{P,w_3} = (1 - \bar{V}_2\rho^\circ) \qquad 2.19$$

where ρ° is the density of the solvent containing w_2 grams of macromolecule and w_3 grams of counter-ions dissolved in w_1 grams of water (constant composition). Experimentally, an apparent specific volume (ϕ') is often determined for situations where the concentration of counter-ions is low and dialysis has brought the solution and solvent to equilibrium. Then, if ϕ' is measured over a range of concentrations and extrapolated to macromolecular infinite dilution, the result is $\bar{V}_2$. If the macromolecule is dissolved in high concentrations of counter-ion (e.g. NaCl, guanidine hydrochloride) and the solution is dialysed to equilibrium, then at equal chemical potentials one measures $\phi_{\mu s}$ which at infinite dilution of the macromolecule becomes

$$\phi_{2,\mu_3} = (\partial V/\partial w_2)_{T,\mu_2}$$

A second condition for defining partial specific volumes in the presence of counter-ions is where the molality of these ions is the same in the solution and dialysate (see Eisenberg, 1976, p. 50, for methods used to reach this state). Then, at infinite dilution of the macromolecule, the partial specific volume is

$$\phi^*_{2,\mu_3} = (\partial V/\partial w_2)_{T,m_3}$$

The value for an apparent specific volume (ϕ') is calculated from the difference in density of the solution (ρ) compared with the solvent (Cassasa and Eisenberg, 1964), using an equation similar to those given in eqs. 1.13 and 1.14, namely

$$\phi' = (1/\rho^\circ)\{1 - [(\rho - \rho^\circ)/c_2]\} \qquad 2.20$$

Creeth and Pain (1967) discuss the precautions required to achieve accuracy in these measurements, while Eisenberg (1976, Chapter 3) discusses the theory. ϕ' is the parameter required for calculating molecular weights from the ultracentrifuge and eqs. 2.15 and 2.16 should be written containing ϕ' in place of $\bar{V}_2$ at finite concentrations of macromolecule. The measurements require accurate estimates of densities, and Lee and Timasheff (1974) were able to achieve a precision of about 0.003 $cm^3\,g^{-1}$ in $\bar{V}_2$ by using the resonant frequency method of Kratky *et al.* (1969) to estimate densities. In Table 2.1 are listed some values of partial specific volumes for solutions of macromolecules in 6 M guanidine hydrochloride. These values show that there are only small differences between $\bar{V}_2$ and the two apparent specific volumes for any one protein. However, there are reasonably large interprotein variations. Often $\bar{V}_2$ for a protein is calculated from the

Table 2.1 True and apparent partial specific volumes ($cm^3 g^{-1}$) of proteins in 6 M guanidine hydrochloride (taken from Lee and Timasheff, 1974, Table 1). $\bar{V}°$ was found by extrapolating the apparent partial specific volumes (ϕ) to infinite dilution of protein; ϕ_{2,w_3} is the apparent partial specific volume at infinite dilution when the molal concentration of 6 M guanidine hydrochloride is identical in solution and solvent; ϕ_{2,μ_3} is the value at infinite dilution of the protein when the chemical potentials of the diffusible components are equal (dialysis). (Reprinted with permission from *Biochemistry* **13** 257 (1974). Copyright by the American Chemical Society)

Protein	$\bar{V}°$	ϕ_{2,μ_3}	ϕ_{2,w_3}
Ribonuclease	0.696 ± 0.001	0.694 ± 0.001	0.694 ± 0.001[a]
Lysozyme	0.702 ± 0.001	0.694 ± 0.001	0.704 ± 0.002[a]
Chymotrypsinogen A	0.733 ± 0.001	0.725 ± 0.002	0.736 ± 0.002
α-Chymotrypsin	0.738 ± 0.001	0.713 ± 0.002	0.732 ± 0.001
Bovine serum albumin	0.735 ± 0.002	0.717 ± 0.001	0.724 ± 0.002[a]
Lactate dehydrogenase (heart)	0.741 ± 0.001	0.736 ± 0.002	0.739 ± 0.001[a]
Catalase	0.730 ± 0.001	0.725 ± 0.002	0.726 ± 0.001[a]
β-Lactoglobulin	0.750 ± 0.002	0.719 ± 0.001	0.728 ± 0.002[a]
Trypsin inhibitor (lima bean)	0.699 ± 0.001	0.698 ± 0.003	0.699 ± 0.001
α-Lactalbumin	0.704 ± 0.001	0.698 ± 0.002	0.701 ± 0.001

[a] Proteins showing dependence of ϕ_{2,w_3} on concentration of the protein.

proportional contributions of the amino acids using the values of partial specific volumes of amino acids tabulated by Cohn and Edsall (1943), and given in Section 2.7.2 because this book is not now universally available. But this calculation makes no allowance for solute–solvent interactions when the solvent contains high concentrations of buffer ions. That is to say, it calculates an apparent $\bar{V}_2$ but not ϕ', ϕ_{2,μ_3} or ϕ_{2,w_3}. Interactions between the ions becomes very important for highly charged macromolecules (Cassasa and Eisenberg, 1964). As can be seen from Table 2.1, even molecules like bovine serum albumin at pH 7 interact strongly with 6 M guanidine hydrochloride, producing a significant change in the value of the partial specific volume. In the native state V_2 is $0.735\ cm^3 g^{-1}$, while in guanidine it changes to $0.727\ cm^3 g^{-1}$ (Reisler and Eisenberg, 1969). If the native value had been used to calculate a centrifugal parameter in an experiment with 6 M guanidine hydrochloride, then the parameter would have been 6.3% too high.

Measuring the relative masses of denatured proteins in solutions of sodium dodecylsulphate raises problems connected closely with those discussed in this section. The detergent has a partial specific volume ($\bar{V}_D$) which differs appreciably from that of proteins and, since it is tightly bound to the proteins, the value chosen for the partial volume of the complex considerably affects the calculated relative molecular mass. If the amount bound was known exactly, the correct partial specific volume could be calculated. But, since this amount is uncertain, special experimental techniques must be devised to measure the relative molecular masses in this solvent. Possibly the best-known method is that given by

Reynolds and Tanford (1976) in which a series of measurements of $M^{app}(1 - \phi'\rho)$ for the protein complex are measured by equilibrium centrifugation in solutions of varying densities. The densities are adjusted by varying the proportions of D_2O to H_2O in the solvent. Thus, when

$$(1 - \bar{V}_D\rho) = 0$$

it is not necessary to know the quantity of bound detergent and the value of $M^{app}(1 - \phi'\rho)$ at this point is that of the protein alone. The procedure works because the density of pure D_2O is greater than the contribution from sodium dodecylsulphate, making it possible to span the zero point with experimental values instead of extrapolating outside the measured values. There will be an exchange of D for H in both the protein and detergent, but this is of secondary importance in the estimate.

Partial specific volume and refractive index increment are two parameters that are functionally related through the Lorentz–Lorenz equation (eq. 2.4, $\bar{V}$ is a function of polarizability; Heller, 1966). Hence, all the comments concerning the three partial specific volumes apply equally to refractive index increments. Practically, the increment required for measuring molecular weights from light scattering (eq. 2.11) is equivalent to ϕ' (eq. 2.20) and determined from the dialysed solution.

Concentration dependence of $S_{20,w}$

Forcing molecules to sediment through a stationary solvent means that work is done on the solvent molecules. Resistance to movement of the solute has three origins: (*a*) frictional resistance of the continuous solvent, (*b*) differences of chemical activity of the solvent in the solution compared with the pure solvent left behind (diffusion), and (*c*) near-neighbour interactions between solute molecules as they sediment. Calculating the centre of mass of the boundary by second moments removes the diffusional effects from the calculation of $S_{20,w}$ (Goldberg, 1953), leaving contributions 1 and 3. Frictional resistance (point (*a*)) is determined by the shape of the molecule and is a minimum for a spherical molecule; this source of resistance remains finite even at infinite dilution. The solute–solute interactions produce the concentration dependence of $S_{20,w}$ which disappears at infinite dilution. Theories which relate this interaction to molecular properties are available (see, for example, Burgers, 1938, 1942; Goldstein and Zimm, 1971; Batchelor, 1972), but unlike the thermodynamic virial coefficients (Chapter 1) these are not derived from a general theory which for simple molecules makes it possible to calculate the coefficients from basic molecular parameters. Observations on the sedimentation of large particles in a fluid show that, as the particle moves, so the fluid flows round the particle in the opposite direction. Similar processes occur with macromolecules suspended in a continuous medium, but here the molecules carry a layer of solvent bound to the surface and this has similar thickness to those of the solvent molecules. This layer may be in equilibrium with the bulk solvent and may be undergoing rapid exchange, but the average mass of solvent carried remains constant. Hence, the volume occupied by the solvated prticle is larger than expected from dry densities (hence the use of apparent specific volumes in eqs. 2.15 and 2.16), while, as the concentration of the macromolecule increases, so the probability that interaction will occur between neighbouring particles also increases. This effect can best be described in terms of excluded volume rather than weight fractions (Burgers, 1942; Goldstein and Zimm, 1971; Batcheler, 1972). Accepting that this model describes concentration dependence suggests that the velocity of a particle is proportional to the available volume of free solvent, which in turn is inversely proportional to the concentration (and volume fraction) occupied by the particle. Practically, it is found for synthetic polymers and nucleic acids that this intuitive model describes sedimentation over a restricted range of concentrations to give the following empirical equation

$$S_{20,\mathrm{w}} = S^{\circ}_{20,\mathrm{w}}/(1 + gc) \qquad 2.21$$

where $S^{\circ}_{20,\mathrm{w}}$ is the sedimentation coefficient extrapolated to infinite dilution and g is an empirical coefficient. Equation 2.21 is not used to describe concentration dependence of sedimentation for globular proteins, but instead a modified version is used:

$$S_{20,\mathrm{w}} = S^{\circ}_{20,\mathrm{w}}\left(1 - \sum_{i=1}^{n} A_i c^i\right) \qquad 2.22$$

where $A_1 = g$ in eq. 2.21; usually only the first term of the sum is required for concentrations less than $10\,\mathrm{g\,l^{-1}}$. Having two empirical equations to describe the same phenomenon is unsatisfactory, although this state of affairs may be acceptable experimentally. Instead of trying to resolve this paradox, workers have concentrated on producing physical and mathematical models to explain the concentration dependence as a phenomenon in its own right. The sources of this dependence lie partly in the excluded volume of the macromolecule and partly in viscosity effects. The effect is more complicated in sedimentation than in diffusion (Chapter 3), because in sedimentation there occurs a mass transfer of solvent in the opposite direction to that of the particle; this does not occur in Brownian motion (see Section 3.3.2 for further discussions on frames of reference). Goldstein and Zimm (1971) produced a realistic model for this dependence by factorizing it into two components, one component being equated with excluded volume and the second arising from solvent–solute and solute–solute interactions. The first component can be calculated from hydrodynamics, assuming the molecular shape is known (this is discussed further in Section 3.3), while the second can be calculated from an energy integral containing the algebraic sum of attractive forces, arising from London–van der Waals' effects, and repulsive coulombic forces. The resulting equations used to calculated this second factor resemble in many respects eq. 1.22, which is used to define the second virial coefficient. In this respect, the Goldstein–Zimm model approaches the van der Waals' equation for non-ideality of gases (eq. 1.21). A less physical approach was made by Rowe (1977) in which he collected the interactive forces together into a volume function which has to be estimated experimentally.

The goldstein–Zimm model is particularly attractive for interpreting concentration dependence, since it provides a theoretical and experimental basis for studying selective solute–solute interactions which are so important biologically in encounter processes between macromolecules, as in long-range recognition of antibodies by antigens (see Section 4.4). In this respect, the concentration dependence of hydrodynamic parameters may be a major way to explore these energetically weak interactions.

Molecular weights estimated from the Svedberg equation

The Svedberg equation (eq. 2.16) was derived for infinite dilution of the macromolecule ($S^{\circ}_{20,w}$ and $D^{\circ}_{20,w}$), but at finite concentrations the model should include concentration dependence of diffusion and sedimention. For a homogeneous solute, the dependence of $D_{20,w}$ on concentration is small (see Chapter 3) and is described by the dependence of relative mass on concentration at equilibrium (Tanford, 1961, p. 380), i.e.

$$D = (RT/F)[1 - c(\partial \ln f/\partial c)_{T,P}] \qquad 2.23$$

where f is the activity coefficient of the macromolecule (see Chapter 1) and F is the molal frictional coefficient. The non-ideal component of eq. 2.23 is given assuming all the concentration dependence of diffusion is derived from the osmotic pressure dependence (see Chapter 3 for further discussion of this), but there should also be a contribution from the frictional dependence. The sign of $\partial \ln f/\partial c$ may be positive or negative depending on the interactions (see Table 1.1). The order of magnitude of this dependence can be similar to that found for sedimentation, but this is not universally true and the two do not compensate in the Svedberg equation. This makes it difficult to calculate accurate molecular masses from velocity experiments at finite concentrations. Heterogeneity of the solute worsens the problem of calculating average molecular masses using eq. 2.16 because the proportionality between $S_{20,w}$, $D_{20,w}$, and molecular masses are different. Graham's law of diffusion for pure gases shows that $D \propto M^{-1/2}$, but this is untrue for solutions because there is an additional viscosity of the solvent to consider. Here, the proportionality can be written as $D \propto M^{-z}$ where z for a sphere is $\frac{1}{3}$ but increases with increasing asymmetry. From the Svedberg equation, it is obvious that $M \propto S_{20,w}\beta$ while M is related to D through the frictional coefficient (Chapter 3). Thus, in a mixture of macromolecules where statistical averaged coefficients are measured, e.g. weight average

$$\bar{D}_{20,w} = \sum_{i=1}^{n} D_i c_i \Big/ \sum_i c_i$$

the ratio of the weight-average diffusion to the sedimentation coefficients does not give a weight-average molecular mass unless $z = \beta = 1$ for all components

(Gibbons, 1971). If however, combining a Z-average diffusion coefficient with a weight average $S^{\circ}_{20,w}$ gives a weight-average molecular mass from the Svedberg equation (at infinite dilution) for all values of z and β (Kinell, 1959).

It is important to realize that a meaningful weight-average sedimentation coefficient can be calculated for any mixture no matter how disperse. This is achieved through using moment analysis to locate the centre of mass of the sedimenting boundaries (for a detailed description of this procedure, see Goldberg, 1953; Schachman, 1959, p. 59; Baldwin, 1957a, b).

Sedimentation equilibrium

The equation for sedimentation equilibrium (eq. 2.15) can be integrated between two radii to give

$$c(r) = c(r_0)\exp\left[YM(r^2 - r_0^2)\right] \tag{2.24}$$

where $c(r_0)$ is the concentration at radius r_0 and

$$Y = (1 - \bar{V}_2\rho_S)\omega^2/2RT$$

This simple solution can be expanded to include conservation of mass within the cell (Rinde, 1929), giving

$$c(\xi) = \frac{c^{\circ}\lambda M \exp(-M\lambda\xi)}{1 - \exp(-M\lambda)} \tag{2.25}$$

where c° is the concentration at rest,

$$\lambda = Y = (r_b^2 - r_a^2)$$

and

$$\xi = \frac{r^2 - r_a^2}{r_b^2 - r_a^2}$$

(r_a and r_b being the respective radii of the meniscus and bottom of the cell). Equation 2.25 is useful when analysing mixtures of non-interacting macromolecules. Scholte (1968) showed that individual concentrations could be estimated from results of a centrifugal experiment with a mixture of synthetic polymers by assuming that the total concentration, $c_T(\xi)$, at any radius of an equilibrated solution equals a linear sum of the individual concentrations, $\Sigma\, c(\xi)$. The mathematical model then becomes a sum of the right-hand sides of eq. 2.25. Thus, providing the number and molecular masses of the species are known, the concentrations of the individuals can be calculated from the experimentally determined $c_T(\xi)$. Naturally, this calculation requires more than one experimental point. In fact, it would be essential to study independently several total concentrations, run at differing speeds and measured over as wide a range of radii as possible. Then, using a suitable statistical routine, the static concentrations could be estimated.

Analysing experimental records

Of equal interest is whether objective molecular masses, concentrations, and number of components in a mixture could be given from such a series of equilibrium concentrations? The answer to this question depends on whether unique solutions can be obtained for fitting a sum of exponentials to a limited range of experimental results. Equations containing sums of exponentials (as in eq. 2.26) often describe physical phenomena, and the statistical problem of fitting such an equation to data of limited precision is well known (if the sum is changed to an integral from zero to infinity then this becomes a Laplacian equation):

$$y(x) = \sum_{i=1}^{p} A_i \exp(B_i X) \qquad 2.26$$

Thus, matching eq. 2.26 to the summed version of eq. 2.25 shows that p would be the number of components, B_i is $\lambda_i M_i$ and A_i becomes $C_i^\circ B_i$. The functional correlation between B_i and A_i when applied to the centrifugal problem means that it is very unlikely that more than two terms could be resolved through the regression procedure. Even if p were known for eq. 2.26, Myhill (1967, 1968) showed that $y(x)$ must be measured with an accuracy which increases as the values of the coefficients (A_i to B_i) become more similar. Normally, only two or three terms can be fitted to experimental records; even then, this is only possible if the coefficients differ by orders of magnitude and if the range of the abscissa is sufficiently wide that it covers significant contributions from all the exponential terms. Analysing natural phenomena through sums of exponentials has always been a thorn in the side of statisticians and new approaches are continually being produced (e.g. Provencher, 1975). Experience shows that as a general rule it is better to combine results from several techniques when applying eq. 2.26 rather than attempting to estimate several parameters fitted to limited information obtained from one experiment.

Experimental considerations

Many experiments are analysed by fitting eq. 2.15b to records in which concentrations are estimated either directly by light absorption or relatively by Rayleigh interference fringes (see Schachman (1959, p. 41) and Lloyd (1974, p. 64) for a description of the otical methods, and Creeth and Pain (1967) for limits on the methods). This procedure appears to be a straightforward statistical problem involving fitting a linear equation to records (from eq. 2.15b, $\ln c$ is linearly related to r^2). If the record does not fit a straight line as determined by statistical tests of significance, then the interpretation of the curvature is not strightforward. Before heterogeneity is assumed to be the underlying cause of the curvature, various experimental perturbations of the record must be excluded. Among the commonest sources of error is the time required to reach equilibrium.

One major disadvantage with equilibrium experiments is the long time taken to reach a quasi-equilibrium state (theoretically, it takes infinite time to reach equilibrium, but practical considerations mean that some reasonable approximation is possible). Theoretical studies (summarized by Fujita, 1961, p. 299) show that the time is proportional to $\Delta r^2/D$ (where Δr is the length of the liquid column) (Mason and Weaver, 1924). The proportionality constant is complex, being functionally related to S and D, but for most practical purposes using short columns it approximates to 0.8. Hence, for a 0.1 cm column of liquid containing a macromolecule having a diffusion coefficient of 10^{-11} m^2s^{-1}, it takes approximately 20 h to reach equilibrium, but doubling the length of the column quadruples the time. Because diffusion coefficients of buffers of relatively low relative molecular masses are at least ten times greater than those of a macromolecule, they reach equilibrium within a few hours. This redistribution helps to stabilize the macromolecular gradient (see Section 2.2.2).

Because of possible thermal degradation of the macromolecules, experiments lasting 20 h at 20°C are not desirable, and attempts have been made to devise procedures which replace equilibrium. Possibly the best known is the Archibald method (Archibald, 1947), where the definition of equilibrium as the state where no nett flux of material occurs is interpreted literally; this pertains for all ultracentrifugal experiments at the meniscus and bottom of the cell. At these two positions for all speeds of the rotor, there always exists quasi-equilibrium. It follows from eq. 2.15a that dc/dr, $c(r)$, and r (where r is that of the meniscus or bottom) are the three ordinates that must be measured. A description of the procedure is given by Schachman (1959, p. 182) while Trautman and Crampton (1959) give a useful analysis of the measurements. Reservations about the accuracy of the Archibald method arose because of difficulties in measuring $(dc/dr)_r$ (using schlieren optics) at the meniscus or bottom. A combination of optical aberrations and rapidly changing dc/dr with r at these positions make it difficult to extrapolate back from the visible part of the dc/dr profile. It would be possible to improve on empirical extrapolation for homogeneous solutes using the Fujita–MacCosham (1959) solution to the Lamm differential equation, but this would bias results if heterogeneity of solute were present. If however, a fluid could be found which was immiscible with water and less dense, but had the same refractive index, then many of the aberrations due to the meniscus would disappear. No such fluid has so far been described, since these three requirements seem to be mutually incompatible. For these reasons, the procedure of high-speed equilibrium experiments (Yphantis, 1964) has displaced the Archibald method, but of course this is a true equilibrium experiment requiring extended time to complete.

The second procedure for hastening the approach to equilibrium is to overspeed the rotor at the beginning of the experiment, so producing an accumulation of material at the bottom of the cell. After a calculated period at this higher speed, the rotor is decelerated to below the equilibrium speed and held there for a short time in order partly to mix the solution before accelerating to the speed of the experiment (Hexner *et al.*, 1961). Howlett and Nicol (1972) examined this idea by simulations, while Richards *et al.* (1968) provide details of the calculations. This method reduces the usual 17 h experiment to about 7 h and, as shown in

Table 2.2 Results from a sedimentation equilibrium experiment with ribonuclease using the low-speed equilibrium method combined with overspeeding (from Richards *et al.*, 1968, Table X). The solvent was 0.1 M NaCl, 0.01 M ethanolamine (pH 9.5); column height 3.2 mm and temperature 25°C. The apparent relative masses were calculated for three parts of the column, namely miniscus (m), bottom (b), and whole column (w). (Reprinted with permission from *Biochemistry* **7** 1054 (1968). Copyright by the American Chemical Society)

Solution	c° (fringes)	$c(b) - c(m)$ (fringes)	$M^{app}_{w,m} \times 10^{-4}$	$M^{app}_{w,b} \times 10^{-4}$	$M^{app}_{w} \times 10^{-4}$
1	28.37	28.86	1.316	1.324	1.321
2[a]	24.16	27.24	1.317	1.328	1.327
3[a]	18.46	21.05	1.338	1.317	1.316
4[b]	14.04	22.78	1.252	1.332	1.316
5[b]	10.80	21.31	1.264	1.324	1.311
6[b]	7.11	14.03	1.239	1.327	1.311
Mean			1.288 ± 0.041	1.325 ± 0.005	1.317 ± 0.06

[a] Centrifuged 27 h (including overspeeding) at 17 250 rev min^{-1}. [b] Centrifuged 10 h (including overspeeding) at 21 740 rev min^{-1}.

Table 2.2, the accuracy is not impaired. The values for M_w and M_Z at all positions of the cell for a 10 h experiment were indistinguishable from values obtained after 27 h, and both were close to the accepted relative mass for ribonuclease. Complementing this procedure are the calculations described by Roark (1976) for determining the time to reach equilibrium when changing speeds of an equilibrated solution.

Discussions on times to reach equilibrium would not be complete without including a warning that all the calculations assume that the transport of the macromolecule is by diffusion and sedimentation. Stirring the solution through mechanical vibrations transmitted to the rotor from the drive or by density inversions caused by inadequate control of temperature would nullify these calculations. Instabilities from these sources become increasingly important as the concentration is decreased. The work of McCallum and Spragg (1972), Triebel (1972), and Fuller Noel and Schumaker (1973) illustrate the errors that can arise from these sources. The effect does not manifest itself as a discontinuity in an otherwise smooth set of results, but forms a smooth continuity from correct to incorrect results. This insidious distortion of results is particularly important when analysing dissociation of macromolecular systems and Figure 2.2 shows a particular case studied by Fuller Noel and Schumaker (1973). Results obtained without the presence of a stabilizing gradient showed an enhanced dissociation of the oligomeric protein which was less when analysed in the presence of a sucrose gradient. This example shows that it is always advisable to test experimental procedures using substances of known relative mass in order to assure oneself that unwanted external factors are not disturbing the experiment. Small

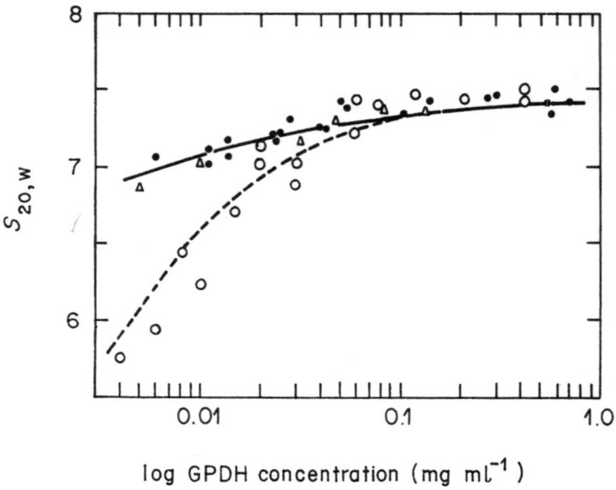

Figure 2.2 The $S_{20,w}$ of native glucose phosphate dehydrogenase as a function of concentration (plotted on logarithmic scale) when determined in stabilized and non-stabilized sedimentation velocity runs at pH 7.0 and 20°C: ○ , non-stabilized results; ●, 0–1%, and ▲, 5–6% sucrose gradients. The solid curve passing through the stabilized values was calculated assuming a reversible dimer–tetramer equilibrium having an association coefficient of $1.8 \times 10^3\ l\,g^{-1}$, while the dashed curve was for unstabilized results using a monomer–tetramer equilibrium having an association coefficient of $5.67 \times 10^8\ l^3\,g^{-3}$. (Taken from Fuller Noel and Schumaker, 1973, *Anal. Biochem.* **41** 1, Figure 1. Reproduced by permission of Academic Press Inc)

temperature inversions occurring during the early stages of the experiment will lengthen the time to reach equilibrium by many times those originally calculated.

2.2 Interpreting ultracentrifugal results

From the limited survey of methods given in Section 2.1 can be judged the reason why ultracentrifugal methods have been so popular when estimating relative molecular masses of medium-sized macromolecules. The instrument is relatively simple to operate, the theory is well defined, and experimental results are not perturbed by irritating artifacts such as dust, etc., as with light scattering. For this reason, the application of the ultracentrifuge is considered in more detail than light scattering or osmotic pressure. Light scattering becomes more useful for studying large macromolecules, but here the theory is complex. In any case, intensity light scattering is slowly being superseded by quasi-elastic light scattering to give diffusion coefficients, and this is considered in detail in Chapter 3.

2.2.1 Non-ideality

Possibly the best recognized problem when calculating relative molecular masses from centrifugal records concerns thermodynamic non-ideality. In Section 2.7.1 is derived the basic sedimentation equilibrium equation for an ideal solution. With non-ideal solutions, this equation is modified to give the non-linear equation

$$M_2 = \frac{\mathrm{d}\ln c}{Y\,\mathrm{d}(r^2)}\left[1 + c\left(\frac{\partial \ln f}{\partial c}\right)_{T,P,c,\ldots}\right] \qquad 2.27$$

Practically, $(\partial \ln f/\partial c)_{T,P,c,\ldots}$ can be expanded to give a mathematical series containing virial coefficients (eq. 1.24) in which all but the first are assumed to be zero, i.e.

$$M_2 = M_{\mathrm{app}}(1 + B_2 M_2 c) \qquad 2.28$$

Since B_2 measures the excluded volume of the macromolecule, it can only be positive (eq. 1.23). In contrast, $(\partial \ln f/\partial c)_{T,P,c,\ldots}$ can be positive or negative depending on solvent conditions for any one macromolecule (see Table 1.1). Thus, the higher-order virial coefficients may be negative or positive depending on the interactions between solvent and solute. Eisenberg (1976, p. 46) discusses the effects that charge has on the virial coefficients (see also Section 1.2.3), but for most globular proteins in moderate ionic strengths (about 0.1 M) the charge on the macromolecule is small and this factor contributes little to non-ideality compared with excluded volume (in contrast with ribonucleotides, Chapter 1).

Usually, globular protein solutions are ideal at acceptable working concentrations but, as eq. 1.23 indicates, the magnitude of B_2 will increase with increasing asymmetry of the molecule. Typical values for B_2 vary from 1.11×10^{-8} l g^{-2} mol for a sphere, to 75×10^{-8} l g^{-2} mol for a rod having $l/d = 100$. Thus, for a sphere of relative mass 10^5 dalton, the estimated mass would be 0.3% low at 1 g l^{-1}, while a similar rod would be 7.5% low at this concentration. Most globular proteins approximate in shape to spheres and have an experimental B_2 of about 3×10^{-8} l g^{-2} mol for relative molecular mass of 50 000 dalton (as given by Munk and Cox, 1972)—but the spread is relatively large, as can be seen by comparing this average with the value for oxyhaemoglobin where B_2 is 5.9×10^{-8} l g^{-2} mol, while cyanohaemoglobin is 3.9×10^{-8} l g^{-2} mol (see Chapter 3 for further discussions of these B_2). This average value shows that if the object of an experiment is to measure an unbiased molecular mass by ignoring non-ideality, then it is necessary to work at concentrations less than 10 g l^{-1}. Introducing this condition also means that the range of concentrations between meniscus and bottom of the cell at centrifugal equilibrium must be restricted, otherwise the apparent masses estimated at the meniscus end of the column will be larger than those calculated at the bottom of the column where the concentration is greatest. In principle, it is possible to make use of this effect to estimate virial coefficients from records of one concentration by generating experimentally a large concentration gradient and then analysing the results to include non-ideality. For this procedure to be successful, it is essential to have smooth accurate records, because the procedure involves numerically differentiating the concentration record in order to calculate the apparent masses at a series of concentrations.

As the macromolecular shape changes from spherical to rod or random coil, then non-ideality contributions become more important. This means that it is always necessary to treat solutions of nucleic acids as non-ideal even when studying concentrations greater than $0.1\,\mathrm{g\,l^{-1}}$. Some experimental compensations occur when working with expanded molecules because it is likely that their refractive index increments will be larger than for the same relative mass wound up to form a sphere from which the solvent is excluded. This makes it possible to examine lower concentrations of nucleic acids by light scattering than is the case with proteins. In any case, the ultracentrifuge is not the most suitable tool for studying molecules as large as many nucleic acids (see Section 2.5), so these molecules are more usefully studied by light scattering. Using light scattering does not remove the problems of non-ideality, but here corrections can only be applied if measurements are made over a range of separate concentrations and angles (see Section 2.12).

2.2.2 Solution densities

Having measured an ultracentrifugal record, the relative masses can only be calculated if the density of the solution and the apparent specific volume of the macromolecule is known (for the latter, see Section 2.1.3). The density of the solution is measured separately from the ultracentrifuge, and it is normally assumed when analysing centrifugal records that the density of the solution is constant with respect to the radius. It is obvious, however, that theoretically this is untrue because the concentrations of macromolecules and buffer molecules increase with radius. An exact formulation of the relationship between molecular mass and changing density for a solution at mechanical equilibrium is complex, but for one at density equilibrium it is readily defined (the result is an equation containing M_2, but this method is relatively inaccurate in determining M_2). Here the macromolecule sediments to a region where $\rho_1 = \rho_2$ and form a gaussian-shaped peak described by

$$c(r) = c_{\mathrm{m},\,0}\exp(-\delta^2/2\sigma^2) \tag{2.29}$$

where $\delta = (r - r_0)$ (r_0 being the radius of the centre of the peak) and

$$\sigma^2 = -RT/M_2\omega^2 r_0^2 I \tag{2.30a}$$

$$I = \frac{\mathrm{d}}{\mathrm{d}r}\left(\frac{\partial \rho}{\partial c_2}\right)_{\mu} \tag{2.30b}$$

In this special isopycnic experiment, a density gradient is generated by the distribution of small molecules in the solvent and the macromolecule accumulates at the radial position where its density equals that of the solvent: the broadening is caused mainly by diffusion and in some part by polydispersity (if present). The conditions of isopycnic equilibrium do not occur in sedimentation equilibrium, so the question remains as to whether it is valid to assume uniform

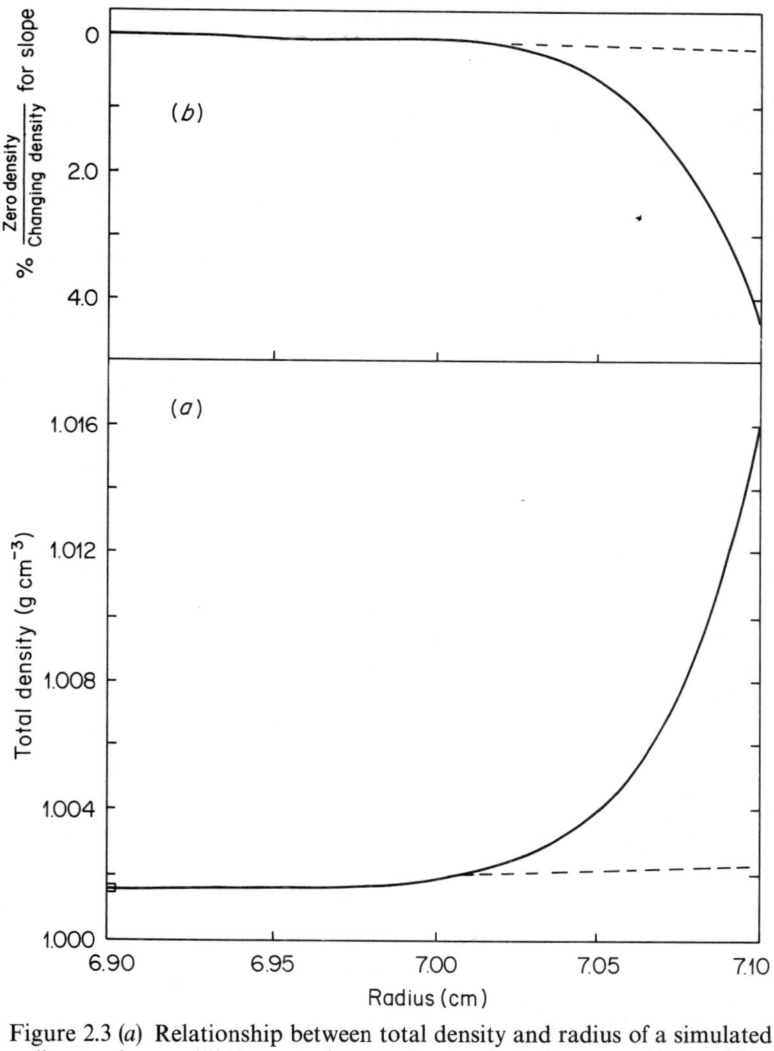

Figure 2.3 (*a*) Relationship between total density and radius of a simulated sedimentation equilibrium run for a molecule of relative mass 66 000 dalton ($\bar{V} = 0.75\ \mathrm{cm^3\,g^{-1}}$) in 0.05 M phosphate buffer pH 7.0 (approximate weight-average relative mass 70): □, 28 000 rev min^{-1}; —, 14 000 rev min^{-1}. Concentration of macromolecule at 6.9 cm was $1.8 \times 10^{-2}\ \mathrm{g\,l^{-1}}$ for 28 000 rev min^{-1} and $3.68 \times 10^{-1}\ \mathrm{g\,l^{-1}}$ at 14 000 rev min^{-1}; the corresponding values at 7.1 cm were $3.4\ \mathrm{g\,l^{-1}}$ and $2.4\ \mathrm{g\,l^{-1}}$. (*b*) The percentage decrease in the experimental slope that would be required to give a constant slope of $\mathrm{d}\ln c$ against $\mathrm{d}(r^2)$ over the whole column

density at sedimentation equilibrium? Simulating an equilibrium experiment for a macromolecule of relative mass 66 000 dalton in a small column of liquid (where the concentration of the macromolecule at the midpoint is equal to the static concentration) shows that the answer depends on the speeds employed. Figure 2.3 shows a simulation for the normal case where the total density increases towards the bottom of the column and high speeds are employed (28 000 rev min^{-1}). The result is an increase in the macromolecular concentration gradient, $\mathrm{d}\ln c/\mathrm{d}(r^2)$, of 4.4% at the bottom compared with the value at the centre of the column (at 7.1 cm). If a constant average density has been assumed in calculating the molecular mass from this experiment, the result would have been an artifact—an increase in relative mass at the bottom of the column. Running at half speed reduces the inaccuracy from assuming constant density, but this is achieved at the sacrifice of the level of gradient in both density and macromolecular concentration, so producing a less stable system. The conclusion is that small increases in relative molecular mass with increasing radius can either be ascribed to redistribution of supporting solutes and macromolecules or an equally possible cause is molecular associations of the type discussed in Chapter 4.

Pressure

Of greater importance than density gradients in the cell are the effects that pressure gradients have on the partial specific volumes of the macromolecules. Considerable pressures can be generated at the bottom of a column of liquid in velocity experiments, and one result of this is to change chemical equilibria between macromolecules. From the thermodynamic definition of partial molar volume, $(\partial V/\partial n_1)_{T,P,n_2}$, can be formed the relationship

$$\left(\frac{\partial \ln K}{\partial P}\right)_{T,n_2} = -\frac{\Delta V}{RT} \qquad 2.31$$

where K is the equilibrium constant for the reaction. Generally, ΔV is small for most reactions involving macromolecules, but occasionally it is sufficiently great that even with moderate pressures (<100 bar) K can alter significantly. Davies and Gutfreund (1976) list typical values of ΔV, and it is interesting to note that the value for the formation of monomers from polymers of myosin shows a comparatively large decrease in volume (-280 cm^3 mol^{-1}). From this, it follows that increasing the external pressure on this system would decrease association to produce a species having a lower relative mass and, consequently, slower sedimentation rate. In the centrifuge, this reversal of the sedimentation can produce instabilities which produce spurious convection boundaries in the cell, as well as distorting the shape of the major boundary. This system has been analysed in detail by Josephs and Harrington (1967, 1968) but, because the effects are complicated by the high concentration dependence of the sedimentation of myosin, it is difficult to obtain clear conclusions. Nevertheless, they were able to show that at speeds of 60 000 rev min^{-1} it is possible to dissociate completely the polymers of myosin at the bottom of the 1 cm column of liquid in an analytical ultracentrifuge. Simulations of these effects show that a large inverted concentration gradient is poduced ahead of the moving boundary and, since negative density gradients cannot occur in experiments, the result is a continual convective stirring in the cell (Johnson *et al.*, 1973). Similar instabilities are given by Schachman (1959, p. 61) for albumin.

A less familiar example occurs when low-density lipoproteins are subjected to pressure (Mauldin and Fisher, 1970) and the result is shown in Figure 2.4. The difference between the two schlieren records given in Figure 2.4 shows that increasing the pressure by increasing the length of the fluid column reduced the proportion of slower-sedimenting macromolecule (these molecules float in this solvent). This pattern would be expected for an associative reaction which gives a negative ΔV value. In sedimentation equilibrium, these effects are not noticeable because the pressures generated seldom exceed a few bars, so including pressure effects in the analysis is not important.

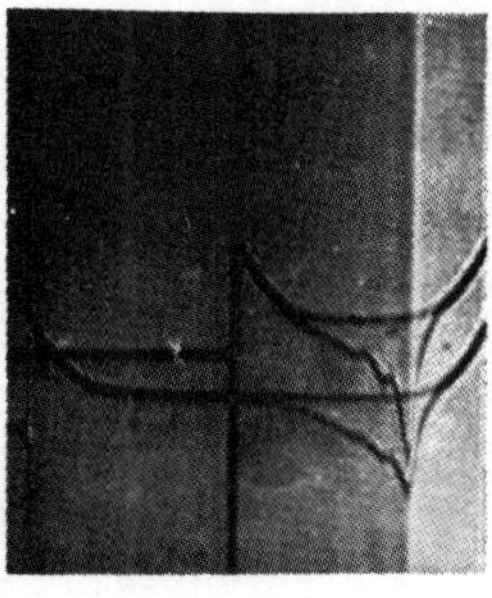

Figure 2.4 Effect of pressure on the ultracentrifugation (flotation) of low-density lipoprotein in KBr solution of density 1.21 g cm^3 (25°C) buffered to pH 5.1. Lipoprotein concentration 12 g l^{-1}. Photograph taken 8 min after reaching the speed (52 640 rev min). Two solutions were run simultaneously, one containing 0.4 cm^3 of solution and the other 0.2 cm^3; the pressure at the bottom of the fuller sector would be about twice that of the less full. The effect of the increased pressure is to reduce the mass of lipoprotein present in the slow-moving component (to right-hand side) and to increase the amount in the faster-moving components, compared with the lower pressure. (Taken from Mauldin and Fisher, 1970, Figure 5. Reprinted with permission from *Biochemistry* **9** 2015. Copyright by the American Chemical Society)

2.2.3 Baselines

Concentrations are usually measured with respect to the solvent (interference and light-absorption optics), but in schlieren optics the distribution of the buffer ions must be measured and subtracted from the macromolecular distribution before a correct distribution of the macromolecule is reached. In both types of record, it is generally assumed that these subtractions are all that is necessary to give unbiased distributions. This appears to be correct when the baselines are recorded at the same time as the solution and when the menisci of both solvent and solution are at equal radii. The importance of this limit can best be appreciated by considering refractometric procedures separately from those involving light absorption.

Refractometric measurements

The most common method in this class is the schlieren system (see Lloyd, 1974, for physical details), which provides a direct measurement of derivatives of refractive index with respect to radius. The records from this method are particularly sensitive to dirt on cell windows, but this can be reduced by using the same window to cover solvent and solution (double-sector cells). Thus, a simple practical solution is possible in this case, but, even if window perturbations are overcome, unequal distributions of buffering ions in the solvent and solution can mean that it is difficult to interpret the resulting records. Most biological molecules are charged, and equating chemical potentials of the solvent by dialysis before centrifuging must introduce unequal concentrations of diffusible ions in the solution relative to the solvent (Donnan effects, Chapter 1; Tanford, 1961, p. 225; Eisenberg, 1976, p. 41). The level of error that this difference produces depends not only on the inequalities of concentrations of the ions but also on their individual refractive-index increments. Normally, the unequal distribution is proportional to the charge and number of moles of polyelectrolyte, both being small for proteins (bovine serum albumin has about 20 net negative charges per molecule at pH 7). With polyelectrolytes such as heparin, where every monomeric unit has a net charge at pH > 2, or nucleic acids at pH 7, then unequal distributions of the counter-ions after dialysis must lead to unequal contributions to the total refractive index of the solution compared with solvent, and this means that baselines cannot be estimated by simple separate measurements. Similarly, when running proteins complexed with detergents (e.g. sodium dodecylsulphate), complications arise which make estimates of baselines difficult. First, the large relative mass of the detergent which complexes with a unit weight of protein (Reynolds and Tanford, 1970) enhances the refractive index of the complex over that of the uncomplexed protein, as well as reducing the concentration of uncomplexed detergent in the solution. Secondly, the solution contains a different proportion of detergent micelles in equilibrium with monomeric detergent than in the solvent. Independent estimates of baselines for these solutions are meaningless and satisfactory corrections can only be made for velocity experiments by extrapolating the baselines in the plateau of the solution back into the regions of the boundary. No satisfactory similar method can be devised for sedimentation equilibrium when using charged macromolecules or detergents because the baseline is not recorded as a separate measurement if interference optics are employed (this is the normally chosen way to record concentration distributions in equilibrium experiments).

Light absorption

Modern analytical ultracentrifuges include photo-electronic scanners for computing absorptions from light transmitted through the solvent and solution (that is, $\log(I_0/I)$), so it would appear that the baseline is included within the calculation. Many inorganic and organic buffers do not absorb light in the spectral regions where biological macromolecules absorb (e.g. 260 and 280 nm) so, unlike refractometric measurements, unequal distributions of ions within the solution do not affect the records. However, light scattered by denatured macromolecules that have adsorbed onto windows, and consequently are not in solution, can produce uncertain baselines. This scattered light gives an apparent increase in light absorbed, and some idea of the magnitude of the problem can be seen from the record shown in Figure 2.5 for ovalbumin (a particularly 'sticky'

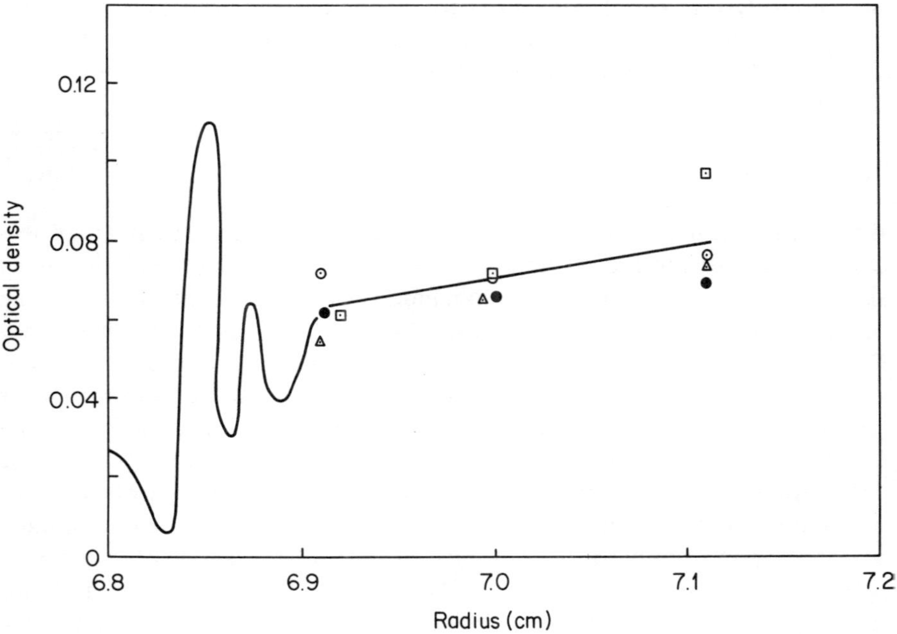

Figure 2.5 Example of baseline formed by denatured protein settling on windows of an ultracentrifugal cell after centrifuging ovalbumin in 0.05 M phosphate buffer, pH 7.0 (absorbance at 280 nm was 0.82) for 18 h at 20°C (13 410 rev min^{-1}). This baseline was recorded after completing the run by removing the protein solution, washing the cell once with buffer, and re-running with buffer in both sectors. Absorbances were measured at 286 nm (●), 280 nm (△), 275 nm (○), and 270 nm (□). The continuous line is the original recording normalized to match the baseline absorbances and showing the meniscus at 6.85 cm followed by Fresnel interference fringes.

protein). This record was collected following removal of the solution from the cell, then replacing it with buffer solution and re-running to give the absorbance at the original equilibrium speed (the original experiment lasted 24 h at 20°C). The fact that the apparent absorption increased with decreasing wavelength according to the $1/\lambda^4$ rule showed that most of the absorption was caused by Rayleigh scattering of light at the windows (see eq. 2.9). Whether this baseline should have been subtracted from the experimental records before calculating relative masses is debatable, but certainly it should be allowed for when estimating concentrations for subsequent calculations of interaction coefficients with associating systems (see Chapter 4).

Most schlieren records are collected on film, but it is possible to use the electronic scanner for this (Reisner and Rowe, 1971; Spragg, 1967). Here, the light transmitted through the solvent is subtracted from that passing through the solution, and the gradients detected by interposing a knife edge between the cell and the scanner at the correct focal plane. With this method, it is assumed that, if no refractive-index gradient were present in the cell, equal intensities of light would be transmitted through solvent and solution. If, however, globules or micelles were present in one sector and not in the other, then light would be lost through scattering and the linearity between the two intensities and refractive-index gradient would be lost. By varying the speeds of the rotor and noting changes in the record, this anomaly can usually be detected but not allowed for in the corrections.

2.3 Relative matrix methods

Absolute physical methods require relatively expensive apparatus and experienced operators for interpreting results. Hence, it is not surprising to find that many short cuts have been explored for obtaining relative molecular masses using simpler equipment. The most successful of these methods employ stationary porous matrices over which the solution is driven either by elution or by electrical potential differences. The macromolecules then separate according to their relative macromolecular volumes (and related to Stokes' radius: Charlwood and Utsumi, 1969) and not according to their relative molecular masses. Since the distribution of sizes of pores in the matrix has been closely controlled during production and these holes are filled with solvent, the proportion of molecules held in the pores compared with those in the free solvent can be described by a partition coefficient which is then related to the molecular volume. The theory for this technique follows that for partition of a solute between two immiscible solvents, and theories of gel chromatography employ the partition coefficient as a molecular parameter (this is a true equilibrium coefficient) for describing the partition between the two phases (Laurant and Killander, 1964; Ackers, 1970, 1975). These models do not apply so readily to gel electrophoresis since here the gel is formed within the tube by polymerizing a monomer (usually acrylamide) using either a chemical catalyst (such as ammonium persulphate) or ultraviolet radiation. The extent of the cross-linking is controlled by the concentration of bifunctional monomer added in relatively small amounts compared with the main monomer. The polymer is a relatively inert matrix which stabilizes the liquid phase against gravitational forces and thermally induced density inversions. Hence, the theory of gel electrophoresis is concerned with relating a molecular parameter to the frictional resistance that the molecule encounters when moving through the covalently linked gel held stationary in an electric field. Both chromatography and electrophoresis require simple equipment for operation. Chromatography utilizes a fraction collector and monitor for estimating concentrations of the macromolecule as it emerges from the column. The electrophoresis technique requires a controllable direct current power supply producing up to a few hundred volts plus stains for detecting the separated components. It is convenient at this stage to describe the two methods as subdivisions of the matrix methods.

2.3.1 Gel chromatography

Extensive surveys of analytical gel chromatography have been made by Ackers (1970, 1975), who describes a general theory in which continuity equations are developed to describe the shape of the bands. In order to appreciate these theories, it is necessary to know that experiments fall into two categories. The commonest application is the one where a band of the macromolecular solution is applied to the top of an equilibrated gel column and this is eluted to give peaks (Figure 2.6a). The second procedure is where a very wide band of the macromo-

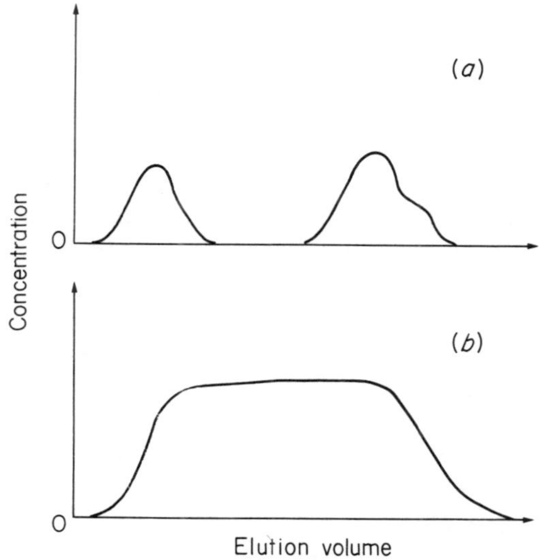

Figure 2.6 Idealized elution profiles from gel chromatography. (*a*) Record obtained from the elution of a thin band added at the start. (*b*) A plateau produced by initially eluting the column with solution containing a constant amount of macromolecule, then following with buffer

lecular solution is applied to the column to produce a plateau of constant composition once the initial boundary has been eluted. This plateau is in true equilibrium with the solution contained within the gel particles. Results from the first technique are useful only for estimating relative molecular masses, while the boundary shapes produced by the second method can be analysed to give molecular coefficients which resemble those for sedimentation, and the analysis employs similar equations to those given in the discussion of velocity sedimentation (Section 2.1.3).

Normally, all that is required to estimate a relative molecular mass using the zonal procedure is to measure the elution volumes of the peaks. Then, after correction of these volumes for the void volume of the column (measured by passing down the column a large molecule which is excluded from the pores), it is found that the logarithm of the Stokes radius of the macromolecule is inversely proportional to this elution volume (Andrews, 1964, 1965). The problem is to discover the proportionality constant between these variables. So far, it has not been possible to calculate this coefficient from a knowledge of the simple properties of the gel (see Ackers, 1975, for details of these analyses), so prior calibration of the column is required using proteins of known Stokes' radii. Generally, this radius is related directly to molecular mass, but this can only be true for macromolecules having the same shape and density (the radius is inversely proportional to the density of the macromolecule or directly proportional to the partial molar volumes). Ward and Arnott (1965) investigated the reliability of the procedure using four glycoproteins of known masses, and showed that there could occur large deviations from correct estimates if the globular proteins were used to prepare calibration curves. Thus, the method is subject to substantial inaccuracy if the shape of the unknown molecule and its constitution are uncertain. Several proposals have been made for including shape within the model, but these require prior knowledge of the geometrical shapes of the pore (Porath, 1963; Squire, 1964; Laurant and Killander, 1964). For these reasons, workers have turned to chromatography of denatured proteins when measuring relative molecular masses, since then it can be assumed that molecules are random coils having radii of gyration based on the Florey–Fox parameters as obtained by Tanford *et al.* (1967) for proteins in 6 M guanidine hydrochloride plus mercaptoethanol. Since Fish *et al.* (1969) derived a relationship between the radius of gyration and molecular mass ($R \propto M^{0.55}$), then the relative mass can be calculated. Although these procedures overcome the objections of shape and density uncertainties, it should be remembered that multisubunit complexes are likely to disintegrate in denaturing solvents, so it is not possible to study complexes by this method.

In the second method of chromatography, the presence of a plateau where the concentration is constant with respect to elution volume makes it possible to provide exact analyses of the shape of the boundaries from the records using a basic transport equation (eq. 2.32) similar in form to the Lamm centrifugal equation (eq. 2.14). As Ackers showed (1970), the total flux (J) through a cross section of the matrix is given by

$$J = \frac{Fc}{\xi} - L\frac{\partial c}{\partial x} \qquad 2.32$$

where F is the uniform flow rate, c is concentration, ξ is the cross-sectional area where partitioning occurs, L is the axial dispersion coefficient (or diffusion coefficient describing boundary spread), and x is the linear coordinate down the column. This equation is essentially Fick's law for diffusion with a moving frame of reference (Gosting, 1956), and the analogy with sedimentation can be seen by putting

$$J = cS\omega^2 r - D(\partial c/\partial r)$$

Analytical solutions to eq. 2.32 are available (Ackers, 1970) and these depend on the definition of a partition coefficient σ ($= V_P/V_I$ where V_P is the penetrable volume accessible to the solute and V_I is the internal volume of the pore accessible to the solvent). The equations include three phenomenological coefficients, namely σ, L, and D, of which σ and D have been defined but L is a new parameter. In fact, the molecular properties that determine L are σ and D, reducing the number of independent variables to two. One important conclusion comes from these theoretical studies, namely the necessity to include concentration dependence in the interpretation of the elution profile. This dependence arises from two sources, σ and D; σ is a hydrodynamic function whereas the diffusion is related to chemical activities of the solute and solvent in the column. This dependence is included through empirical equations similar to those used in sedimentation (eqs. 2.21 and 2.22) by replacing S with elution volumes. In single solutes, this dependence produces boundary sharpening in the trailing edge (largest elution volume) and spreading in the leading edge (Winzor and Scheraga, 1963, 1964).

Possibly the most powerful application of analytical gel chromatography is not in determining relative molecular masses but in detecting and measuring effects produced by differing conformational changes in the molecule. In this technique, the native macromolecule is applied to the column and, after it has moved into the matrix, the treated molecule is layered on top and trails behind the first. The two-component-thick zone is eluted by buffer. At the interface between the two macromolecular solutions, there occurs either an accumulation or loss of material depending on whether the treatment has increased or decreased Stokes' radius. Hummel and Dryer (1962) were the first to employ this technique and since then it has been improved considerably (Gilbert *et al.*, 1973). The power of this technique lies in its sensitivity to detect and measure small changes in molecular conformations and interactions. Figure 2.7 (reproduced from Gilbert *et al.*, 1973) shows an elution profile produced for oxyhaemoglobin layered on methaemoglobin in a column of polyacrylamide. For practical purposes, these haemoglobins have the same relative molecular masses but differ in the energy of interaction between the two dimers that form the tetramer (see Chapter 4). The

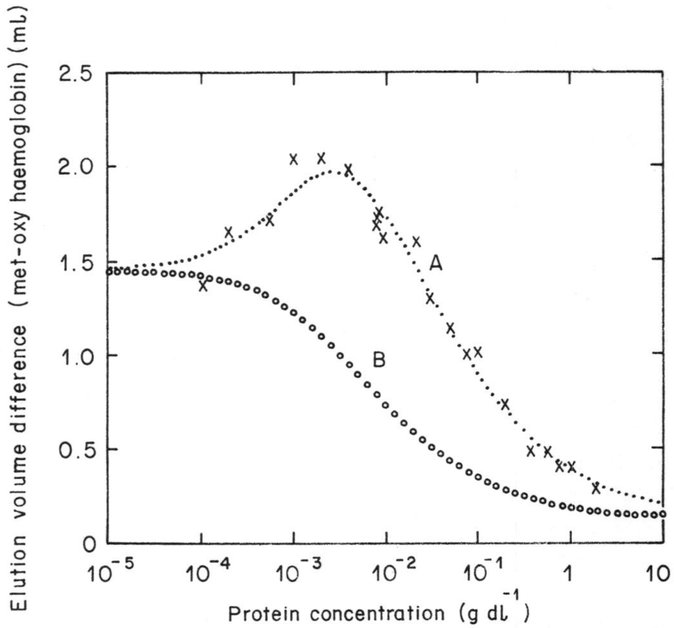

Figure 2.7 Profile obtained in difference chromatography of human oxyhaemoglobin and methaemoglobin at 5°C, pH 7.0 in 0.1 M Na^+ (0.09 M Cl′ + phosphate) with 10^{-3} M EDTA. Two associations are involved

$$2(\alpha\beta)^{oxy} \rightleftharpoons (\alpha\beta)_2^{oxy}$$

and

$$2(\alpha\beta)^{met} \rightleftharpoons (\alpha\beta)_2^{met}$$

Curve A is the line fitted by least-squares analysis to the experimental points for the ratio of coefficients

$$L_{2,4}^{met}/L_{2,4}^{oxy} = 0.62 \pm 0.01$$

giving a free-energy difference of 1.11 kJ mol^{-1}. Curve B is a similar line drawn for

$$L_{2,4}^{met}/L_{2,4}^{oxy} = 1$$

for comparison. (From Gilbert *et al.*, 1973, *Nature, New Biology* **235** 110, Figure 1. Reproduced by permission of Macmillan (Journals) Ltd)

protein concentrations employed in the experiment (Figure 2.7) were 80 mg l^{-1} and the profile recorded by light absorption at the isobeistic point for the two proteins. From the volume of the perturbation in the plateau, a difference in association constant was calculated and was equal to 1.11 kJ mol^{-1}, which is less than RT at 0°C. This small difference could not be measured by subtracting results from separate experiments using the ultracentrifuge (for example), especially at this concentration of protein. Incidentally, these results also show complete hybridization between the subunits of the two proteins. Thus, using this technique, unambiguous comparisons of macromolecular sizes can be made between a reference macromolecule and an unknown.

A discussion of gel chromatography would not be complete without mentioning the direct scanning procedure of Ackers (discussed by Ackers, 1970). Most applications of gel chromatography estimate concentrations of the macromolecules as they elute from the column, but Ackers was able to measure light absorbances of macromolecules in the column and follow the distribution of concentration within the matrix. To make this procedure useful it is necessary to estimate baselines accurately and, since a column of Sephadex (cross-linked dextran) 1 cm in diameter has an absorbance of about 1.8 at 280 nm, nearly all the stray light must be removed from the system. The procedure provides information on the true partition coefficients within the column. Unfortunately, the need to produce scanners, monochromators, and recording equipment increases the complexity of the equipment and decreases the simplicity of gel chromatography, so this method has not been widely applied.

2.3.2 Gel electrophoresis

Electrophoresis has been neglected in this book mainly because free-boundary electrophoresis has many limitations and its popularity has declined over the years. Both technical and theoretical factors (summarized in Bier, 1959) have meant that it has been replaced by gel electrophoresis. This powerful tool has superseded many of the classical methods for detecting heterogeneity, and within limits it can provide reasonable estimates of relative molecular masses of subunits. Generally, polyacrylamide is the supporting matrix, although polysaccharides and porous glass beads have been used: the macromolecular solution is applied as a band at one end before electrophoresis is begun. Detection after separation is by selective staining of the macromolecule still in the gel. Staining is usually irreversible, so the method can only be used as an analytical procedure. Elliot (1976) has improved detection by employing optical methods to locate bands. These methods can be employed while separation is in progress, making the procedure analogous to the apparatus used in free-boundary electrophoresis. Careful preparation of the gels is important for optical recording, since inhomogeneities in the gel would be recorded as bands in the record.

In common with all electrophoretic methods, the velocity of a non-conducting macromolecule (v) in an electric field (E) measured in V cm^{-1} depends on both charge (Q) and frictional coefficient (f) of the molecule (eq. 2.33), namely

$$v/E = Q/f \tag{2.33}$$

Equation 2.33 is of little use for analysing gel electrophoresis because not only is the path length in the gel uncertain (so v cannot be measured) but also one cannot be sure whether the field strength is uniform along the length of the gel column.

The main reason for quoting eq. 2.33 is that it does remind one that the property of the macromolecule that varies with its size and which electrophoresis measures is Q/f, and of this ratio the charge is determined not only by the constitution of the native macromolecule but also by the pH of the solution. Further, because $f \propto S^{-1}$ (see Section 2.1.3), there will be a concentration dependence of f (and hence the size) which is similar in magnitude to that discussed for the sedimentation coefficient. Neglecting concentration dependence makes it possible to use gel electrophoresis for determining relative masses providing Q could be made proportional to mass and the shape could be made the same for all macromolecules. To produce a proportional Q means that the charge per amino acid residue (or nucleotide) must be constant for all proteins (or nucleic acids). This condition is more likely to be met with nucleic acids than proteins, but under the correct conditions (heated in high concentrations of detergent) most proteins will absorb sodium dodecylsulphate in a constant weight of detergent per weight of protein (Reynolds and Tanford, 1970). The result is a polymer in which the number of added charges introduced from the detergent increases linearly with increasing relative mass, and this generally overwhelms the natural charge of the protein. Denaturation of the proteins by the detergent also reduces the shapes of many to those approaching a random coil and so gives approximately uniform frictional coefficients. Shapiro *et al.* (1967) first demonstrated that this property could be enlisted in determining relative masses, while Weber and Osborn (1969) and Dunker and Rueckert (1969) showed it to be accurate to within 5–10%. The experimental procedure depends on measuring in the same gel the mobility of the unknown molecule (v_k) relative to marker protein complexes of known molecular masses, when $v_k \propto (\log M)^{-1}$. As with gel chromatography, the accuracy depends on there being a constant relationship between the frictional coefficients, charge, and relative molecular mass for both the markers and unknowns. Hence, the most likely source of inconsistency is the inability to reduce the native protein to a random coil with the detergent. There are reports of anomalous results for cross-linked proteins (Davies and Stark, 1970; Griffith, 1972) caused by bonding restrictions imposed through the predetermined amino acid sequences; this endows the molecules with more rigidity than is normal and they retain remanents of their original shapes even in relatively high concentrations of detergent. Reduction of disulphide groups can reduce this source of uncertainty for most globular proteins. Attempts have been made, however, to allow for this lack of randomness by measuring v_k in a series of gel concentrations when the limiting mobility (v_0) at zero gel concentration can be found from

$$v_k = v_0 \exp(-K_r T) \qquad 2.34$$

where K_r is an empirical retardation coefficient and T is the total weight of monomeric acrylamide plus *bis*-acrylamide in the gel (Furgusen, 1964). From the experimental plot of $\log v_k$ against T can be estimated K_r, and this is assumed to be the constant which is related to the relative mass. Repeating this plot for a range of standards will calibrate the gel. However, as Dunker and Kenyon (1976) show, even this modification does not detect unusual conformations, so the inevitable conclusion is that although gel electrophoresis can provide useful estimates of relative masses, the accuracies are limited and unpredictable.

2.4 Applications

Known erroneous theories have been omitted from the previous discussions but, even when the theory appears to have been correctly applied, false interpretations of data can still occur. This section attempts to give a cautionary summary of some of the interpretative difficulties found in the literature and which arise from the meaurement of relative masses under widely ranging conditions. Possibly the overriding question when measuring relative masses is: what is the limit on accuracy that can be expected in determinations of relative molecular masses of macromolecules? The answer determines the confidence placed on differences between experiments. Experience shows that the methods are capable of high accuracy but the limit is set by the purity of the sample; thus, discussions of accuracy must include descriptions of how heterogeneity distorts the measurements. A further factor is whether the magnitude of the mass influences the proportional accuracy, and whether some methods are better suited to studying large particles while others are restricted in their range. These factors are considered in the following sections.

2.4.1 Heterogeneity and accuracy

The work of Svedberg and his group (summarized by Svedberg and Pederson, 1940) showed that haemoglobin was a discrete molecule. In the work, they used concentrations which by present standards would be regarded as high (greater than $10\,\mathrm{g\,l^{-1}}$). This was fortunate because haemoglobin dissociates to give significant proportions of dimer at $1\,\mathrm{g\,l^{-1}}$ (see Chapter 4). The ignorance of the Swedish workers about the extent of this dissociation was an advantage, since they were able to demonstrate by working in high concentrations that biological macromolecules had reproducible molecular masses. Only in later work was concern expressed about the possibility that contamination will distort results, and one major practical problem that all the earlier workers had to solve was to devise methods for purifying macromolecules from mixtures of naturally occurring macromolecules. Conventional fractionation procedures used at the time were not satisfactory for macromolecules; fractional distillation was obviously useless, while although fractional crystallization was possible, it was difficult to crystallize all macromolecules. Of the remaining methods, only fractional precipitation in varying concentrations of salts proved reliable for proteins. It is interesting to note that it took a further 25 or more years before procedures were developed for fractionating nucleic acids with sufficient resolution that a mixture could be resolved into many discrete molecules. Indirectly, these problems of purification meant that much of the early knowledge about the physical chemistry of proteins was gained from investigations on serum albumin and haemoglobin; these were the two proteins that could be readily prepared relatively purely in quantity. Tissue proteins could only be extracted after homogenization of the tissues in a suitable solvent followed by fractionation of this complex mixture. This route made them less attractive to physical chemists. Nevertheless, the work with albumin and haemoglobin provided many guides to future macromolecular studies, and, although the principles derived from this work have been absorbed into biochemical methods, some of the premises and cautions have been lost. Some of these supporting pillars are discussed here.

Relative mass of albumin

It is interesting to find that as late as 1960 Phelps and Putnam stated that, 'the molecular weight of bovine serum albumin cannot be stated better than as being within the range 65 000 and 70 000. . .', even though the first estimates were made in 1927! This represents an uncertainty of over 7% caused not by random errors (which is precision) but by variations between laboratories using various techniques. Hindsight shows that many workers had taken the necessary precautions to obtain accurate estimates of the molecular masses, but these values were obscured by the large number of inaccurate estimates reported in the literature. For example, in 1954 Gutfreund, using osmotic pressure, obtained an estimate of 66 000 ± 1000 dalton for bovine serum albumin (Gutfreund, 1954). This accuracy was achieved by careful purification of the sample as well as by extending the measurements down to sufficiently low concentrations that he was able to extrapolate with certainty beyond the regions of non-ideality. This example is one of the few accurate estimates up to 1960, so it is not surprising that Phelps and Putnam (1960) made this plea. Finally, Squire *et al.* (1968) laid the problem to rest by showing that early high estimates at low concentrations were caused by contamination. Using centrifugal equilibrium and carefully purified samples, Squire *et al.* (1968) estimated the relative molecular mass of bovine serum albumin to be 66 700 ± 400 dalton, agreeing well with Gutfreund's earlier estimate and with the value obtained by sequencing of albumin (Brown, 1975). There are other estimates of ultracentrifugal parameters for bovine serum albumin which precede Squire's work and allow for heterogeneity, but these seem to have been ignored when evaluating relative mass. Lessons can be learnt by retracing some of these early experiments made with bovine serum albumin.

The early estimates of molecular mass of bovine serum albumin were generally made by combining estimates of sedimentation with diffusion coefficients (Svedberg equation, eq. 2.16). A careful study made by Harrington *et al.* (1956) on bovine serum albumin at pH 1.9 showed that increasing the ionic strength from 0.1 M to 0.3 M raised $S_{20,w}$ from 0.34 ps to 0.37 ps. A further increase occurred when the pH was raised to 5.0 at ionic strength 0.1 M to give $S_{20,w}$ of 0.44 ps. Using similar conditions, Champagne and Sadron (1954) showed that diffusion increased in the same manner as $S_{20,w}$, showing that the molecule remained intact but changed shape as the conditions were changed ($\bar{V}$ variations are small over the range of pH used, but could change with ionic strength, so part of the changes found with altered ionic strength may be due to using a constant $\bar{V}$ in all the calculations). Historically, these observations showed that the earlier suggestion that albumin dissociates into subunits (Weber, 1953, using fluorescence depolarization) was incorrect. The stability of albumin was confirmed by light scattering (Harrington *et al.*, 1956) which, incidentally, gave a relative molecular mass of $72\,000 \pm 2000$ dalton. These analyses were made allowing for concentration dependence and the results are one of the 'accurate' estimates of that time. Baldwin (1957a, b) was one of the first physical investigators to attempt to incorporate paucidispersity in his analysis of ultracentrifuge parameters. To do this, Baldwin made accurate measurements of $S_{20,\mathrm{w}}$ over a range of concentrations, employing second-moment analysis to locate the centre of mass of the sedimenting boundary. This calculation had not been made by earlier workers. The range of concentrations studied and his accuracy were such that he could detect curvature of the $S_{20,\mathrm{w}}$ versus concentration curve (cf. eq. 2.22) and produced the following equation to describe the dependency (c is measured here in g %):

$$S_{20,\mathrm{w}} = 0.443_8(1 - 0.054c - 0.004c^2) \qquad 2.35$$

(Baldwin normalized to 25°C, but this has been changed to 20°C in eq. 2.35 using known densities and viscosities of water). Combining $S^\circ_{20,\mathrm{w}}$ (0.443_8 ps) with the diffusion coefficient ($6.97 \times 10^{-11}\ \mathrm{m^2\,s^{-1}}$: Champagne and Sadron, 1954) gave the relative mass of 66 500 dalton for the major component in the mixture. Using the independent method of ultracentrifugal equilibrium, Squire *et al.* (1968) confirmed this result.

Quantifying heterogeneity

Retrospectively, we know that Baldwin's estimate was correct, yet he used similar preparations of albumin as his contemporaries (Squire *et al.*, 1968, removed the dimer and higher polymers by gel chromatography, a procedure not available in Baldwin's time), so how was he able to achieve the accuracy he did with this impure sample? First, he began the study by assuming that his preparation was impure and he developed a method that not only detected the proportion of impurity but also went some way to allowing for its presence in the estimates of $S_{20,\mathrm{w}}$. To do this, he defined a parameter for heterogeneity

$$p^2 = \sum_i \alpha_i (S^\circ_i - \bar{S}^\circ_i)^2$$

where

$$\alpha_i = c^\circ_i \Big/ \sum_j c_j$$

(the fractional amount of component i in the mixture) and

$$\bar{S}^\circ = \sum_i \alpha_i S^\circ_i$$

the average $\bar{S}^\circ$ (the subscripts 20,w have been omitted to improve clarity in these relationships). In order to calculate p, the temporal movement of the centre of mass and the broadening of the boundary must be measured. Then, from an analysis of the dependence of moments with time, Baldwin (1957a, b) gave the following definition of p in terms of the measured variables (omitting the dependence of diffusion on concentration):

$$p = [Z/\omega^4 r_{\mathrm{m}}^2 F(t)]^{1/2} \qquad 2.36a$$

$$Z = \sigma^2 - \sigma_0^2 - 2\omega^2 \bar{S} \int_{t_0}^{t} \sigma^2 \, \mathrm{d}t \qquad 2.36b$$

$$\bar{S}\dagger = \bar{S}^\circ - a_1 c - a_2 c \text{ (as in eq. 2.22)} \qquad 2.36c$$

$$F(t) = (t^2 - t_0^2) + \tfrac{4}{3}(\omega^2 \bar{S})(t^3 - t_0^3) \qquad 2.36d$$

Here σ_0^2 is σ^2 at zero time, t_0; σ^2 is the variance of the schlieren profile, which was assumed to be gaussian and estimated from $[A/H\sqrt{(2\pi)}]^2$, where A is the area and H is the maximum height of the schlieren (derivative) curve. In order to carry out the necessary calculations, it is essential to measure several coordinates across each record of the boundary. Modern instrumental developments make this operation a trivial job and it seems sensible to apply Baldwin's procedure routinely when analysing ultracentrifugal velocity experiments. Further, the definition of p makes it directly comparable with the dispersions given in eq. 3.34 for diffusion.

Since σ is a function of diffusion (eq. 2.36b), Z allows for diffusional spreading of the boundary, while p estimates the non-linear dependence of the velocity on time caused by paucidispersity (p has dimensions of time). Estimates of p for two samples of albumin showed that the paucidispersity varied between 0.069 and 0.1 ps; this is equivalent to between 15.6 and 22.6% contamination. Having shown that the samples of albumin were paucidisperse and having estimated the proportional contamination, Baldwin was able to estimate the $S_{20,\mathrm{w}}$ and $D_{20,\mathrm{w}}$ of the main component to give eventually an accurate relative mass.

Workers not familiar with the ultracentrifuge may still wonder why the dimer did not separate from the monomer during the velocity experiment because these two are not reversibly associated. The reason is that diffusional spread of the boundaries reduces resolution in the radial direction. Low levels of contamination (less than 25% by weight) by macromolecules of similar sedimentation rates—i.e. $S(\text{dimer}) = S(\text{monomer}) \times 2^{2/3}$—are not always visible by simply viewing the photograph of the schlieren profile. The detection requires analysis similar to that given by Baldwin. The usable radial path length is less than 1 cm in analytical ultracentrifuges and longer lengths are necessary to separate molecules having similar sedimentation rates. Of course, with mixtures containing two or more components of similar masses, the shape of the boundary becomes noticeably distorted after prolonged centrifuging.

Baldwin's work was concerned solely with records of derivative curves (schlieren optics), but similar equations can be used to analyse records of concentration profiles (Trautman *et al.*, 1969). In all these analyses, it is essential to know the number of components in the mixture, and this must be a limited number, if the accurate $S_{20,w}$ of the main component is to be calculated. Other methods make it possible to detect contamination by small amounts of a large number of components. First, consider the case of the sedimentation of a single macromolecule in a sector-shaped cell when the concentration of the macromolecule in the plateau (c_p, measured directly as in light absorption or indirectly by integrating the derivative curve) is related to the sedimentation by the radial dilution law for a sector-shaped cell, namely

$$dc/dt = -2\omega^2 Sc \qquad 2.37a$$

$$c_p(t) = c_p(0)\exp(-2\omega^2 t) \qquad 2.37b$$

(eq. 2.37 is derived by putting $\partial c/\partial r = \partial^2 c/\partial r^2 = 0$ in eq. 2.14). Thus the linear dependence of $\ln c_p(t)$ on time (t) should give a slope which when divided by $-2\omega^2$ gives S. Naturally, S can also be estimated from the sedimentation of the boundary, so the two estimates should be identical for a pure sample (within statistical errors). If the two estimates do not agree and the plateau value of S is larger than the boundary value of S, then the sample must have contained small amounts of heavier molecules. If the impurities had smaller S than the main component, then the calculated S from the plateau would be less than that from the boundary. These calculations can only give quantitative estimates of heterogeneity if the number and sedimentation coefficients of the components are known. This is because, in a mixture of non-interacting macromolecules, the total concentration is

$$c_t = \sum_i c_i$$

and from eq. 2.37b

$$\sum_i c_i \Big/ \sum_i c_p(0) \simeq c_t \Big/ \sum_i c_p(t) \simeq \exp\left(-2\omega^2 t \sum_i S_i\right)$$

Thus, plotting $\ln c_p(t)$ against time will produce a straight line within the errors, having a slope for impure samples proportional to an average S. This is a sensitive test for detecting small amounts of impurities having a wide range of sedimentation coefficients, and shows that S calculated by this procedure is not a reliable estimate of the S of the major component. The calculations for estimating the plateau S can be coupled with the calculations of moments for the boundary because, in both, the concentrations need not be in absolute units (the calculations of moments include estimates of the area of the derivative curve and, hence, the concentration of the macromolecules in the plateau).

Sensitivity of equilibrium methods to heterogeneity

So far, the discussions of accuracy have concentrated on results for albumin from the ultracentrifuge, but other methods were employed to determine the relative mass during this period of investigation. Why were these unable to provide more accurate estimates before Baldwin (1957a, b)? It seems that other methods are less successful than the centrifuge in detecting heterogeneity. For example, the light-scattering experiments carried out by Harrington *et al.* (1956) gave an inaccurate relative mass (72 000 dalton) for albumin without indicating heterogeneity, yet the estimate of the second virial coefficient proved to be relatively accurate. The sensitivity of light scattering to heterogeneity is low because no temporal separation of components can occur. So, unless there is massive contamination (and a weight-average relative mass of 72 000 dalton corresponds to 6.7% contamination by dimer), these experiments do not readily detect heterogeneity. If heterogeneity is suspected, and both the number and molecular masses of all the species are known, then the weight-average molecular mass can be used to calculate the relative proportions of the components. Measuring the scattering at different angles over a range of concentrations does produce a distorted Zimm plot for heterogeneous solute but, since these plots are generally not sensitive detectors for the presence of relatively small macromolecules in the presence of large molecules, this mode of detecting heterogeneity is not routinely applied.

The insensitivity of light scattering to heterogeneity is one example of the Achilles heel of all equilibrium methods; ultracentrifugal equilibrium suffers in the same way. At equilibrium in the centrifuge, as much as 10% dimer of a relative mass similar to that of dimerized albumin would go undetected in an equilibrium experiment. If, however, the relative mass of the molecule were already known, then even this small amount of contamination could be resolved. Figure 2.8 illustrates this point using simulated data from sedimentation equilibrium experiments run at two speeds, i.e. curve A, at 14 000 rev min^{-1}, and curve B at 28 000 rev min^{-1}, and corresponding with the results given in Figure 2.3. The simulations are for a molecule of relative mass 66 000 dalton contaminated with 10% (w/w) dimer, and the total concentrations at equilibrium were calculated from eq. 2.15. Even using the exact concentrations of the simulation, the data for the mixture fitted a straight line (as for eq. 2.15a) at both speeds when the test for curvature was made at the 95% probability level. From the slopes of this straight line were calculated relative masses of 71 164 dalton (Figure 2.8a) and 67 671 dalton (Figure 2.8b). The true weight-average relative mass for this mixture was 72 600 dalton, showing that the slopes from both speeds did not equate with the expected weight-average relative mass, although approached it at the lower speed. Rounding errors in the arithmetic produced a systematic error of between 500 and 900 dalton and are responsible for the failure of reasonable statistical tests to detect curvature. Random errors in a real experiment would exceed these arithmetical errors, making it impossible to detect heterogeneity. Hence, it is not surprising that statistical tests applied at a single speed are unreliable proofs of homogeneity; the test may be improved by analysing results from the same solution run at several speeds.

The concentrations assumed to apply in the simulation given in Figure 2.8 were sufficiently low that the solution could be considered ideal, but the question of what happens when non-ideal solutions are studied by equilibrium methods is of greater practical interest. Fortunately, an answer is available for this case, because Munk and Cox (1972) have simulated it by choosing a mixture

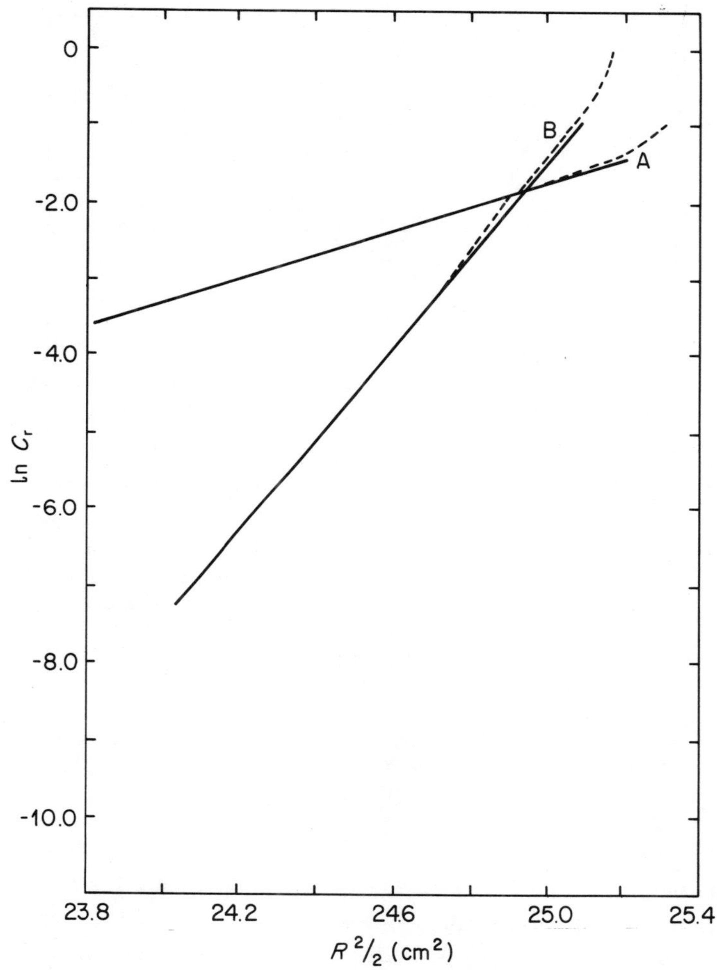

Figure 2.8 Simulated results for the relationship between concentration (ln c_T) and radius ($R^2/2$) for a mixture of two non-interacting macromolecules of relative masses 66 000 and 132 000 dalton mixed in the respective proportions 9:1 (static total concentration $= 1 \text{g} \text{l}^{-1}$). The continuous lines are those fitted by regression analysis, and the dotted lines are the true data. The simulations were made for speeds of 14 000 rev min^{-1} (A) and 28 000 rev min^{-1} (B)

containing monomer of relative mass 36 000 dalton contaminated with dimer in various proportions, but all dissolved in 6 M guanidine hydrochloride. In the simulation, they included a contribution from a second virial coefficient ($\equiv B_2 + (\bar{V}/M_2)$, where $\bar{V} = 0.73 \text{ cm}^3 \text{ g}^{-1}$ and $B_2 = 1.8 \times 10^{-8} \text{ l mol g}^{-2}$ to give $6.16 \times 10^{-7} \text{ mol g}^{-2}$). As with the ideal case, the plots of ln c against r^2 (Figure 2.9) gave relatively good straight lines which would not show significant curvature when errors are added as in real experiments. Furthermore, it was surprising to

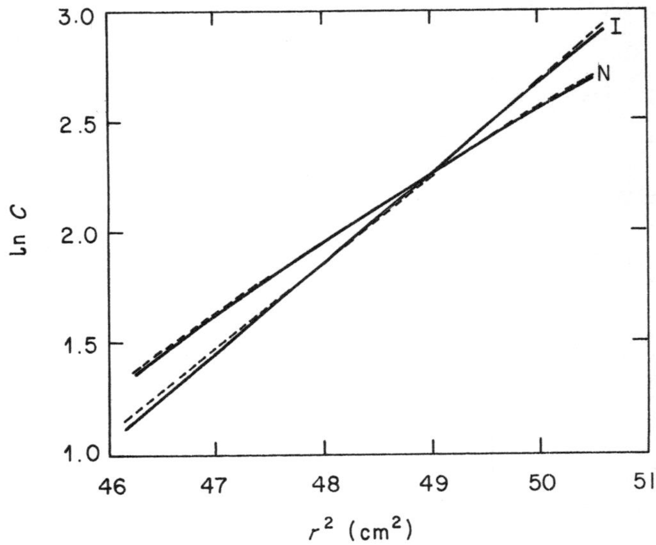

Figure 2.9 Simulated results for a mixture of two macromolecules in which the apparent molecular masses (*M*app) have been calculated for ideal (I) and non-ideal (N) conditions. Full lines are the curves expected for homogeneous solute, and the broken lines are for the heterogeneous solute. The concentration is in Rayleigh fringes (10 fringes represents 34 g l^{-1}). (From Munk and Cox, 1972, Figure 3. Reprinted with permission from *Biochemistry* **11** 687. Copyright by the American Chemical Society)

find that the plot of the estimated $1/M_{app}$ (calculated from the slopes) against concentration also gave a straight line which again masked the presence of a second component, and the slope of this line produced an estimate of B_2 which was very similar to the original. These simulations demonstrate how unreliable equilibrium measurements are in detecting heterogeneity and how non-ideality can mask heterogeneity. Only if the number of components and their relative masses are known can the results from equilibrium be used to study heterogeneity.

The simulations given in Figure 2.8 illustrate one further point about interpreting equilibrium results for heterogeneous solutes, namely how unreliable the static concentrations are when calculated from the midpoint of the profile. In earlier years, this pivot point was used to interpret equilibrium results (Yphantis, 1960) when absolute concentrations were required from records collected using Rayleigh fringes. For the results shown in Figure 2.8, the total concentration at the calculated midpoint ($\Delta r^2/2$) from the straight line and using total concentrations was 19% lower than the true static concentration; the presence of the dimer has distorted the profile by this amount.

Moment analysis for heterogeneity

If fitting a simple model (eq. 2.15) to equilibrated data does not detect heterogeneity, then the model must be changed. One sensitive method for detecting impurities is to compare the various averages of relative molecular mass; these are number-average (M_n), weight-average (M_w), and Z-average (M_Z). In a pure sample, these are all equal, but impurities distort the higher averages relative to the lower averages. The three averages can be calculated from a single ultracentrifugal equilibrium experiment using the definition of moments (σ_i; Fujita, 1961) as described by Roark and Yphantis (1969). The logarithmic relationship between concentration and square of the radius (eq. 2.15) gives M_w from the gradients at each radial ordinate (eq. 2.38a), and from this can be calculated the other two averages using the following relationships:

$$\sigma_w(r) = \mathrm{d} \ln c/\mathrm{d}(r)^2 \qquad 2.38a$$

$$\sigma_n(r) = \frac{c(r)}{\langle 1/\sigma_w(r) \rangle \mathrm{d}c(r)} + \frac{c(\mathrm{m})}{\sigma_n(\mathrm{m})} \qquad 2.38b$$

$$\sigma_z(r) = \sigma_w^2 \frac{\mathrm{d}(1/c\sigma_w)}{\mathrm{d}(1/c)} \qquad 2.38c$$

$$\sigma_i(r) \equiv M_i(1 - \bar{V}_i \rho_S)\omega^2/2RT \qquad 2.38d$$

where the integration constant, $c(\mathrm{m})/\sigma_n(\mathrm{m})$, in eq. 2.38b contains the ratio of the meniscus concentration and its number-average moment, $\sigma_n(\mathrm{m})$. As many values of each moment can be calculated as there are measured coordinates (but minus one), and heterogeneity can be estimated from the ratios of these moments at each concentration.

Unfortunately, random errors are magnified when experimental records are differentiated (as when calculating $\sigma_Z(r)$, eq. 2.38c). Thus, a 1% error in σ_w produces 3% in σ_Z, making it necessary to smooth the record first by fitting a polynomial (preferably orthogonal) to the data. From the empirical coefficients produced by this fit can be calculated the integrals and derivatives. The number average σ_n is a special case (eq. 2.38b), since its calculation requires a knowledge of $\sigma_n(\mathrm{m})$ in order to calculate the integration constant, and this is not available from ultracentrifugal records. When high speeds are employed, $c(\mathrm{m})$ is sufficiently low that the ratio $c(\mathrm{m})/\sigma_n(\mathrm{m})$ can be ignored, but at lower speeds, where this is not true, it becomes necessary to assume $\sigma_w(\mathrm{m}) = \sigma_n(\mathrm{m})$.

It is questionable whether calculating moments would be sufficiently accurate that the level of contamination simulated for Figure 2.8 would be detected. At a radius of 7.088 cm, where the contribution from the dimer is greatest, the ratio σ_Z/σ_w was 1.0023 at 28 000 rev min^{-1} and 1.0014 at 14 000 rev min^{-1}. These values would be within experimental error, confirming the view that only if values for relative molecular masses are already known can low-level contamination be detected by equilibrium experiments. Other methods are available for studying heterogeneity using the ultracentrifugal equilibrium method (a summary is given by Wan and Adams, 1976), but individuals should test a chosen method experimentally using synthetic mixtures of known composition before relying on the results.

The lesson to be learnt from this survey of equilibrium methods is that it is wise to seek more than one method when estimating relative masses. Possibly the best tool to complement absolute physical methods is gel electrophoresis, where the proteins are not only separated into well resolved components but their relative movements can be used to calculate approximate relative masses. This method can be used to monitor purifications, and when all the impurities have been removed (remembering that many active units are oligomeric and exist as an associating system; see Chapter 4) then accurate estimates of relative masses can be made by one of the physical methods. The widespread use of gel electrophoresis makes it worth exploring the lower limits of sensitivity of this method. If bands are detected by staining, then the background residual strain must be reduced to almost zero if a 10% contamination is to be detected. Even if high concentrations of the mixture are applied at the beginning (say 1% w/v protein), then diffusion during the experiment would reduce the maximum concentration to less than 0.5% in any band; this reduces the concentration of the 10% contamination to 0.05% w/v. This concentration would be difficult to detect by staining methods, but is especially bad for a protein dissolved in a gel which has a variable baseline. Thus, it is possible to be fooled into thinking that a preparation is clean because insufficient material was applied to the gel (Dr N. G. Anderson has called this 'purification by dilution', a slightly cynical term but conveying the problem). Further, since staining is often species-specific (as with coomassie blue for proteins), contamination by other macromolecular types (e.g. carbohydrates) in the solution would not be detected by simple staining. These contaminants would contribute to refractometric measurements, so they would give positive tests for heterogeneity in light scattering and some ultracentrifugal experiments but not in gel electrophoresis. For this reason, it should not be surprising if in tests of purity the results from gel electrophoresis disagree with refractometric methods. A marked improvement in gel electrophoresis is possible if, when detecting bands, refractometric detection (Elliot, 1976) is combined with staining.

Comparative techniques

One approach that can be more sensitive in detecting changes in samples than any of the methods discussed so far is the differential method for transport experiments (see Figure 2.6). These experiments can be analysed with great sensitivity if the experimental profiles of the boundary are compared with those calculated using well defined theories. This method is of particular interest in centrifugal velocity experiments where the shape of a sedimenting boundary can be calculated from the Lamm equation (eq. 2.14). To make this comparison useful, the integration of the differential equation must include all the known molecular parameters including concentration dependence. Two procedures have been developed for this numerical integration; one by Dishon *et al.* (1966) which depends on a mathematical matrix procedure to solve the difference equations at each radius, and the other by Cox (1971) using a finite interval to calculate fluxes at hypothetical interfaces within the cell. These integrations require lengthy calculations and have only been used to indicate the shape of curves produced in a variety of experimental situations. It is possible to extend the calculations by iterating between the computations and the experimental records until a point is reached at which no further improvement can be achieved in the comparison. Then, if a single component had been assumed in the calculations, any deviations at the leading and trailing parts of the experimental curve must be attributed to contamination. To make this a sensible analysis, the experimental record must be collected with precision and true allowances made for non-linear baselines, optical aberrations, and uncertain timing of records (see Chapter 4 for further details of results). Approximate calculations of this type were made by Spragg *et al.* (1976) to test an 'on-line' computer system when analysing impure preparations of albumin. Here, an integral solution of the Lamm equation (Fujita and MacCosham, 1959) was used to calculate the expected profile. In addition to providing pictorial comparisons, this analysis provided estimates of S and D for one of the components with the minimum of assumptions concerning shapes of the profiles (McCallum and Spragg, 1972).

2.4.2 Effect of baselines

Discussions of heterogeneity and accuracy for results from hydrodynamic measurements show that, although the first priority must be to assume heterogeneity in the interpretation of experimental records, the results that emerge need not be unambiguous. However, before these tests can be applied, it is essential to reduce to insignificant levels the systematic errors in the records (random errors must also be reduced, but a reasonable level can be tolerated). The sources of these systematic errors varies with the experimental method. For example, in light scattering, the light scattered from dust or by flaring round the cuvette walls are sources of unwanted intensities which must be reduced or allowed for in the analysis (for example, by measuring at varying angles with solvent in the cell). This compares with the ultracentrifuge, where inhomogeneities between cell windows produce spurious baselines when using schlieren or interference optics (these factors are discussed in Section 2.2.3). These errors can be reduced by careful experimental procedures which include measuring optical imbalance between sectors when no macromolecule is present. Then providing these perturbations are small, the bias in estimated relative masses is small at high concentrations (greater than $1\ g l^{-1}$). At lower concentrations, corrections for baselines produce greater distortion of the final answer. Some indication of the effect that correcting for baselines has on ultracentrifugal results can be seen from results shown in Figure 2.10 taken from experiments using absorption optics. In one case, removing a positive uniform baseline of absorbance 0.05

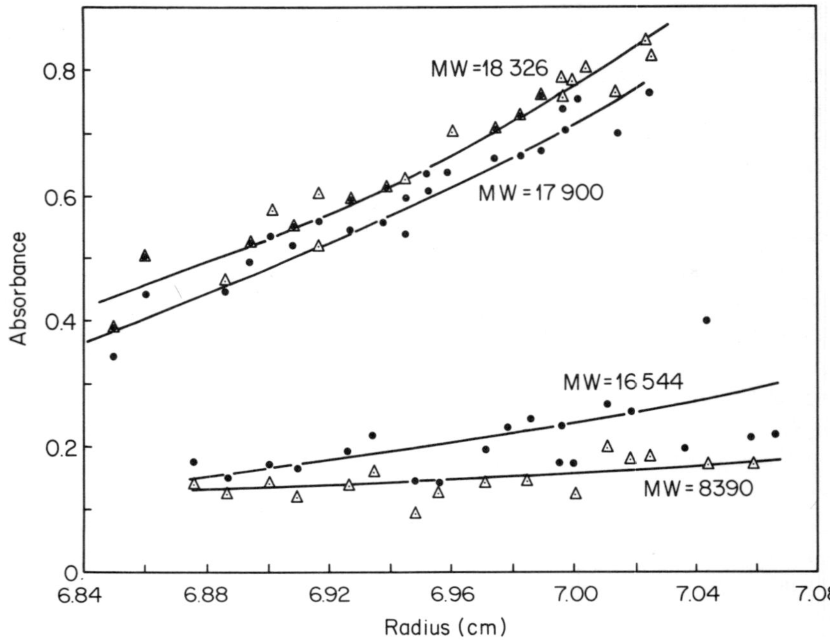

Figure 2.10 The effect that correcting for non-zero baselines has on the estimated relative masses of two concentrations of myoglobin in an ultracentrifugal equilibrium experiment (speed about 17 000 rev min^{-1}): Δ, uncorrected for baseline; ●, corrected for baselines, estimated at the end of the run. The absorbances were measured at 287 nm. A discerning worker would not normally accept the result for low absorbances despite the apparent success of the correction for these data

(calculated using natural logarithms) decreased the estimated relative mass for myoglobin from 18 326 dalton to 17 900 dalton, while, in another case, removing a linear negative baseline of similar magnitude for a run at lower concentrations increased the relative mass from 8390 dalton to 16 544 dalton: this large change would not be acceptable normally. Horse myoglobin has a relative mass of 16 500 dalton calculated by amino acid composition, but even the average of the two estimates would be 17 222 dalton, making this approximately 4% higher than the truth. So far, no universal recipe appears to exist for unambiguously estimating baselines in the ultracentrifuge. One method that has been used is to remove the solution at the end of the experiment, replace it with solvent, and remeasure. If adsorption of macromolecules has occurred on windows, the result is a non-linear baseline of the form shown in Figure 2.5. Although the absolute value of the absorbance is small, it would have major effects on the corrected slopes at low concentrations. Another procedure is to overspeed the rotor at the end of the experiment in order to pack the solute to the bottom of the cell, remembering that the higher species will distort the profile formed from small solutes in the solution compared with that occurring during the equilibrium. The only general recommendation is to try various procedures with molecules of known relative mass, and to compare these results with expected values.

With light scattering, it is not possible to allow for errors in baselines, and the cure is to prevent dust accumulating in the solution by using micropore filters or to design cuvettes so that flaring of light is reduced to a minimum. Correction through making separate experiments are possible, but are of dubious reliability, especially when correcting for dust, since the separation of macromolecules from the dust will depend upon the relative masses and shapes of the dust particles compared with the macromolecule in solution.

2.5 Estimating relative molecular masses of large aggregates

So far, discussions of the accurate estimation of relative masses has largely been concerned with macromolecules in the range of mass 10 000 dalton to about 200 000 dalton. Many active biological units are formed from an ensemble of these macromolecules, and accurate estimates of the relative masses of both the subunits and aggregates are required when studying these aggregates. Normally, the same techniques are used to study large particles as those for smaller macromolecules. However, it is now necessary to adjust experimental conditions to allow for the increased size. Naturally, the sources of error discussed earlier apply equally when studying large aggregates, but additional problems arise which are unique to large aggregates.

When using ultracentrifugal equilibrium to estimate relative masses, it is necessary to run at low speeds. Thus, to reach equilibrium in a 2 mm column with $c(\mathrm{m})/c(\mathrm{b}) = 3 \times 10^{-3}$ requires a speed of 28 000 rev min^{-1} for serum albumin, 780 rev min^{-1} for a relative mass of 10^7 dalton, and 250 rev min^{-1} for 10^8 dalton. These are maximum speeds because it is possible with absorption optics to use smaller concentration differences along the column. Although these low speeds can be achieved with all analytical ultracentrifuges, practical experience shows that precession of the rotor about its vertical axis at low speeds introduces uncertainties in the measurements of radial ordinate which thus make it unrealistic to work at low speeds. This precession occurs at all speeds, but the lateral movement is greater at low speeds. Even at 20 000 rev min^{-1}, Williams (1976) finds that a rotor can precess by about 10^{-2} mm, which represents a large relative movement when dealing with 2 mm columns of liquid. At all speeds, the concentrations measured at discrete radii are really averages over a small but finite radial dimension, and the extent of this averaging increases as the speed is reduced.

Light scattering from large particles

Since the analytical ultracentrifuge used in the equilibrium mode is not accurate for large particles, then experimentalists must turn to light scattering. Equation 2.10 was derived for light scattered from a particle having particulate dimensions which were small relative to the wavelength of the incident light. With molecules having dimensions greater than 10nm, this can no longer be assumed, and allowance must be made for internal interference of the scattered wavelets coming from oscillators separated spatially in the particle. As van Holde (1971, Figure 9.2) and Tanford(1961, p. 297) demonstrate, interference reduces the apparent volume of the particle calculated using eq. 2.10. The extent of this reduction can be calculated using

$$P(\theta) = \frac{1}{N^2} \sum_{i=1}^{N} \sum_{j=1}^{N} \frac{\sin hR_{ij}}{hR_{ij}} \qquad 2.39$$

where $P(\theta)$ is the scattered intensity at θ relative to that expected from a small particle where no internal interference occurs. R_{ij} is the distance between any of the N pairs of oscillators i and j (at the limit, atoms may be used here, but generally only groups of condensed monomers produce sufficient effect for practical purposes), and h is the wave vector

$$h = \frac{4\pi n \sin(\theta/2)}{\lambda}$$

The values of R_{ij} cannot be determined from an experiment nor calculated unless the spatial coordinates of a dissolved macromolecule are known. Thus, eq. 2.39 is largely of theoretical interest, and a procedure is employed for extrapolating experimental observations into a region where eq. 2.10 can be applied. This region occurs when either $h = 0$ or $R_{ij} = 0$. The first condition occurs where $\theta = 0$ ($\sin\theta = 0$, that is, forward scattering), while the second becomes small relative to h at either long wavelengths or for small particles. Detection of scattering of long wavelengths (radiofrequency) is considerably less sensitive than for visible light because the photon energies are so much smaller and attenuation by the solvent becomes important. Hence, the procedure normally followed is to measure scattering at a range of angles and concentrations, then extrapolate the records to zero angle and concentration—Zimm plot (Zimm, 1948) and discussed in more detail by Phelps (1974) and Timasheff and Townend (1970). Accuracy with this procedure can only be achieved if scattering measurements are made down to small angles where the greatest change in $P(\theta)$ occurs. The Zimm plot is constructed by plotting $\Omega c/R$ (eq. 2.10) for each angle against $\sin^2(\theta/2) + ic$ where i is an arbitrary scaling constant required to expand the abscissa (10^3 is often chosen, but any value of this magnitude would work). The result is a grid of points which can be connected in the manner shown in Figure 2.11. The line for $\theta = 0$ is constructed from the knowledge that at this angle the abscissa has a value of Ωc. Extrapolating this line to zero abscissa gives $1/M_2$ and the slope equals $2B_2$, the second virial coefficient. Connecting the points of varying concentration but equal scattering angle produces the horizontal lines, and these can be extrapolated to zero concentration (when the abscissa is $\sin^2(\theta/2)$) to give the line for $c = 0$. From this is estimated values for $\Omega c/R$ when $c \to 0$, and the initial slope of this line (if the plot is curved) is a measure of the radius of gyration of the particle. When polydisperse samples are examined by this procedure, the result is a weight-

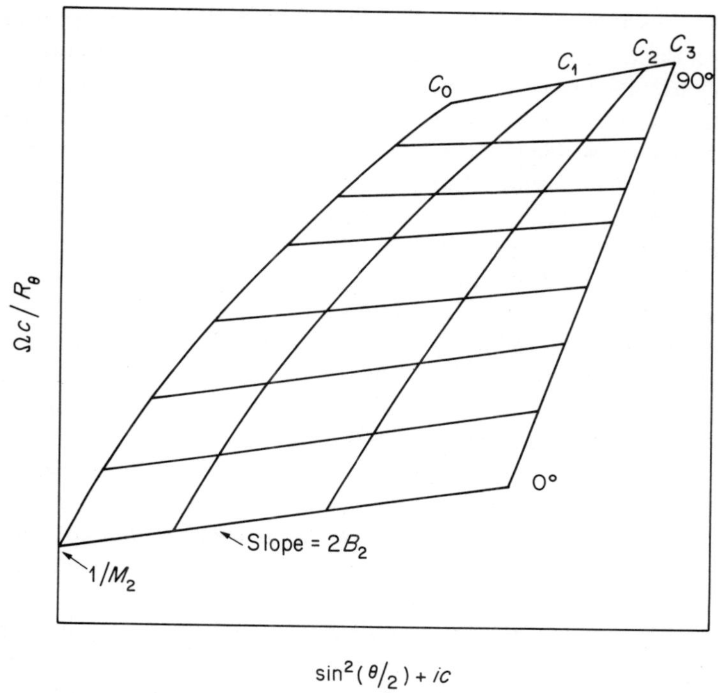

Figure 2.11 An idealized example of a Zimm plot for a polydisperse macromolecule

average relative molecular mass, but the value for the virial coefficient and radius of gyration are complicated averages.

The Zimm plot has proved particularly valuable for studying nucleic acids, where the expanded molecules have large radii of gyration compared with globular proteins (for DNA, see Doty and Bunce, 1952). For example, DNA (deoxyribonucleic acid) with a relative molecular mass of 4×10^6 dalton has a radius of gyration of 117 nm, but if the same mass had been packed tightly, as in a globular protein, then the equivalent radius of gration would have been about 15 nm. Hence, when studying molecules having expanded dimensions, or when associated to form large aggregates, it must be assumed that interference of the scattered light will occur and the analyses must include a Zimm plot.

In order to maintain accuracy as the size of the particle increases, it is necessary to extend measurements further towards zero angle. Practically, this lays great stress on the design of the equipment, and usually the limit of measurement is about 10° relative to the incident beam. This, in turn, limits the accuracy for scattering from very large particles. Furthermore, this need to extend measurements near to zero angle emphasizes the attention required to remove dust from the solutions, because large particles contribute disproportionally to the forward (low-angle) scattering (see Timasheff and Townend, 1970, for a full discussion on experimental procedures). Heterogeneous samples containing wide-ranging molecular sizes are readily detected by the Zimm plot, but numerical problems occur with mixtures when extrapolating to zero angle and concentration.

If, instead of employing temporal-averaged intensities of scattered light (elastic scattering) for studying large particles, quasi-elastic scattering is employed, then it is possible to measure accurately the diffusion coefficients of large particles (see Chapter 3). Combining these measurements with $S^{\circ}_{20,w}$ estimated from velocity experiments with the ultracentrifuge (although equilibrium measurements are difficult with the ultracentrifuge, velocity measurements can be made readily with large particles) gives accurate relative masses of large particles. This involves measuring both coefficients over a range of concentrations and extrapolating back to infinite dilution. The value of g, the concentration-dependence term for the ultracentrifuge (eq. 2.21, expressing concentration as a volume fraction) for large particles, will be the same for large aggregates as for small macromolecules having the same shape and charge. But, since the opportunity to form asymmetric bodies is greater for larger particles than with small, a wide range of values for the coefficient is recorded for aggregates. For example, the concentration dependence for tobacco mosaic virus (having an axial ratio of about 0.05) is about $0.023\,\mathrm{l\,g^{-1}}$, which compares with $0.005\,\mathrm{l\,g^{-1}}$ for serum albumin (axial ratio >0.5; Baldwin, 1957a, b). It should be mentioned that gross asymmetry of a particle is not the sole factor in determining the value of concentration dependence, because occasionally conglomerates are found which are known to be spherical from electron microscopy but have abnormally high concentration dependence in the centrifuge (pyruvate dehydrogenase complex; Danson *et al.*, 1979). This high dependence can be explained by assuming that the particle is constructed like a sponge and filled with holes to give a volume calculated from overall molecular coordinates which is not consistent with that calculated from the mass and density of proteins. It would be difficult to envisage a similar situation occurring with large single-species macromolecules. These factors stress the need to extrapolate centrifugal results back to infinite dilution before calculating relative molecular masses, as well as indicating the sort of molecular picture that can sometimes be drawn from the results. Generally, the dependence of diffusion on concentration is negligible for neutral molecules or charged molecules which have been fully shielded by the inclusion of suitable salt concentrations (see Chapter 3). Thus, at finite concentrations, the value for the relative mass calculated by combining these two coefficients ($S_{20,w}$ and $D_{20,w}$) could be appreciably lower than that measured at infinite dilution. Pusey (1973) compared relative masses obtained for viruses by four separate procedures and showed that the S/D method is probably more reliable than any of the others. He assessed the error to be less than $\pm 5\%$ by this method if the diffusion was measured by fluctuation spectroscopy (Chapter 3).

2.6 Lessons to be learnt

History has provided examples of both successful and unsuccessful interpretations of estimates of relative masses in solution. The work discussed in this chapter shows that it may be irrelevant to ask what is the relative molecular mass of a biological macromolecule, since what may be more important is how does the composition of the solution affect the activity and, hence, the apparent mass. Even molecules which are known to be active as monomers can aggregate when in high concentrations (see Chapter 4). Accepting this premise does not reduce the responsibility of the experimentalist to ensure that estimates are precise, accurate, and refer to the macromolecule under discussion. Serious investigations may produce trivial results in the final analysis, because the correct physical model was not used to calculate parameters or because interpretations did not allow for errors in the analysis of the record. Possibly the most important lesson to be learnt from these discussions is the importance of considering the solution as a whole entity and not experimentally to treat the macromolecule as if it were an inert mass in a continuous medium. This overall conclusion forms the underlying theme through the five itemized summary points given below.

(1) *Purity of sample*

Samples of biological macromolecules must always contain impurities, but the important question is: what are the proportions and types of impurity? With molecules like bovine serum albumin, very little effort is required to separate polymers from the monomer, but, with other molecules, slow reactions may continually frustrate attempts to purify the molecule. For example, oxyhaemoglobin (Fe^{2+}) is readily oxidized to methaemoglobin (Fe^{3+}) even under mild conditions. Not only does this affect the light absorption of the molecule but conformational changes induced by the oxidation may modify the interactive properties. Haemoglobin is a specialized molecule, but its example emphasizes the need to monitor for undesirable side-effects caused by hydrolysis or partial degradation of molecules during their preparation. Very large macromolecules like nucleic acids are particularly susceptible to breakage by shear forces produced by mixing or reaction with free radicals of oxygen in alkaline solutions. Tests showing that biological activity is maintained throughout the investigation are useful, but in the absence of a knowledge of absolute reaction rates these can only be relative tests, and it is wise to employ other physical tests to show purity. For example, techniques that include conservation of mass as a parameter in the analysis can monitor for loss of material during an experiment; also, using two or more techniques that measure different but related properties of the macromolecule should yield the same final result (e.g. refractometric and light-absorption measurements to measure concentration).

(2) *Interaction with small diffusible ions*

When analysing results from a solution containing high concentrations of diffusible ions, it should always be assumed that these ions interact with the macromolecule to give changes in hydration or electrostriction of the solvation shell. Practically, this is reflected in changes of properties such as partial specific volume or refractive index measurements of the macromolecule.

(3) *Concentration ranges*

A single estimate at one concentration is seldom sufficient to define the relative molecular mass of an active complex; it certainly does not demonstrate purity unless the records are precise and can be analysed in detail. Measurements over a range of concentrations provide information from which virial coefficients and specific interactions can be calculated (Chapters 3 and 4). The virial coefficient gives a qualitative view of the molecular volume plus insights into long-range interactions (explored further in Chapter 3), while the specific interactions provide insights into its biological activity. The lower limit of the range will be determined by which techniques are employed to monitor the concentrations, but, even if this can be extended down to a very low limit, other instrumental factors may cloud the interpretation. Transport boundaries are particularly susceptible to thermal instabilities at low concentrations. Each technique has its limits at both ends of the concentration range, and these must be assessed through standards before embarking on detailed analyses of results from new macromolecules.

(4) *Calibration of the method*

When employing absolute procedures like osmotic pressure, light scattering, and the ultracentrifuge, it would seem unnecessary to calibrate the procedure before examining an unknown system. However, if only for the reasons outlined in points (1), (2), and (3), it can be reassuring to obtain accepted masses by using a proven procedure for well defined macromolecules when subjected to identical conditions to those used in studying the new system. Here, the choice of the standard is important because, almost without exception, natural macromolecules have some undesirable property when studied outside the standard range of conditions. For many macromolecules, ill defined associations and dissociations are known to occur, while at high concentrations the values for the virial coefficients may not be known accurately. A molecule like myoglobin is a useful standard because its relative mass at infinite dilution is known accurately from amino acid analyses and sequences, while its partial specific volume is also known. Furthermore, it has a high light-absorption coefficient, making it easy to monitor results at low concentrations. However, this molecule is small (relative molecular mass about 17 600 dalton, but this has species variations), making it unsuitable as a standard when studying large aggregates. Bovine serum albumin can be purified to provide large quantities of a relatively stable standard of relative mass 66 500 dalton, and again its amino acid composition and sequence are known. This could provide a second standard, but there are few larger macromolecules whose solution properties are well defined until one reaches TMV (tobacco mosaic virus).

(5) *Differential methods*

When absolute measurements are secondary to a comparative study, then differential procedures are the best mode of analysis. Systematic errors are reduced by this method as compared with experiments in which two absolute numbers are subtracted one from another. Here, the interpretations of the results may pose problems.

2.7 Appendix to Chapter 2

2.7.1 Derivation of the Svedberg equation

(The basis of this section is Svedberg and Pederson, 1940.) For a binary solution at chemical equilibrium, the chemical potential of the solute (μ_2) is given by

$$\mu_2 - \tfrac{1}{2}M_2\omega^2r^2 = \text{constant} \qquad 2.40$$

where $\frac{1}{2}M_2\omega^2r^2$ is the centrifugal potential. Differentiating eq. 2.40 gives

$$\mathrm{d}\mu_2 - M_2\omega^2r\,\mathrm{d}r = 0 \qquad 2.41$$

We know that

$$\mathrm{d}\mu_2 = \left(\frac{\partial\mu_2}{\partial P}\right)_{T,n}\mathrm{d}P + \left(\frac{\partial\mu_2}{\partial n_2}\right)_{T,P,n_1}\mathrm{d}n_2 + \left(\frac{\partial\mu_2}{\partial n_1}\right)_{T,P,n_2}\mathrm{d}n_1$$

and

$$\left(\frac{\partial\mu}{\partial P_\delta}\right)_{T,n_2} = \bar{V}_1(\text{solvent})$$

Substituting this into eq. 2.41 gives (without some subscripts)

$$\bar{V}_1\,\mathrm{d}P + \left(\frac{\partial\mu_2}{\partial n_2}\right)\mathrm{d}n_2 + \left(\frac{\partial\mu_1}{\partial n_2}\right)\mathrm{d}n_1 - M_2\omega^2r\,\mathrm{d}r = 0 \qquad 2.42$$

From the Gibbs–Duhem relationship we get

$$\sum_{j=1}^{i} n_i\left(\frac{\partial\mu_i}{\partial n_j}\right)\mathrm{d}n_j = 0$$

for two components ($i \neq j$). Therefore, multiplying eq. 2.42 by n_i and summing for $i = 1, 2$, we get

$$\sum_i n_i\bar{V}_i\,\mathrm{d}P - \sum_i n_iM_i\omega^2r\,\mathrm{d}r = 0 \qquad 2.43$$

Substituting for dP and rearranging gives

$$\sum_i\left(\frac{\partial\mu_2}{\partial n_i}\right)_{T,P}\mathrm{d}n_i - \left(M_2 + \bar{V}_2\frac{\Sigma_i n_iM_i}{\Sigma_i n_i\bar{V}_i}\right)\omega^2r\,\mathrm{d}r = 0 \qquad 2.44$$

and this is the basic equation for sedimentation equilibrium. For an ideal solution, this becomes

$$\sum_i \left(\frac{\partial \mu_2}{\partial n_i}\right)_{T,P} \mathrm{d}n_i = RT\,\mathrm{d}\ln n_2 \qquad 2.45$$

which, when substituted into eq. 2.44, gives

$$RT\,\mathrm{d}\ln n_2 - \left(M_2 + \bar{V}_2 \frac{\Sigma_i n_i M_i}{\Sigma_i n_i \bar{V}_i}\right)\omega^2 r = 0 \qquad 2.46$$

Now

$$\frac{\Sigma_i n_i M_i}{\Sigma_i n_i \bar{V}_i} = \rho_\mathrm{S}$$

the density of the solution at r, and thus introducing this into eq. 2.45 gives

$$M_2 = \frac{RT(\mathrm{d}\ln n_2)/\mathrm{d}r}{(1 - \bar{V}_2\rho_\mathrm{S})\omega^2 r} = \frac{RT\,\mathrm{d}n_2/dr}{(1 - \bar{V}_2\rho_\mathrm{S})\omega^2 r n_2} \qquad 2.47$$

Normally, concentration is used in place of a mole fraction to give the usual equation for sedimentation equilibrium:

$$M_2 = \frac{RT\,\mathrm{d}c_2/\mathrm{d}r}{(1 - \bar{V}_2\rho_\mathrm{S})\omega^2 r c_2} \qquad 2.48a$$

$$M_2 = \frac{2RT\,\mathrm{d}\ln c}{(1 - V_2\rho_\mathrm{S})\omega^2\,\mathrm{d}(r^2)} \qquad 2.48b$$

Returning to the Lamm differential equation (eq. 2.14), then at equilibrium $\partial c/\partial t = 0$ and

$$\left(\frac{\partial c_2}{\partial r}\right)_{T,P} = \frac{M_2(1 - \bar{V}_2\rho_\mathrm{S})\omega^2 rc}{RT} \qquad 2.49$$

while

$$\frac{\partial^2 c}{\partial r^2} = Kc + Kr\frac{\partial c}{\partial r} = Kc(1 + Kr^2) \qquad 2.50$$

where

$$K \equiv \frac{M_2(1 - \bar{V}_2\rho_\mathrm{S})}{RT}$$

Introducing 2.49 and 2.50 into the Lamm equation (eq. 2.14) gives

$$M_2 = \frac{RT}{(1 - \bar{V}_2\rho_\mathrm{S})}\frac{S_2^\circ}{D_2^\circ} \qquad 2.51$$

which is the Svedberg equation relating S° to D° to relative molecular mass for velocity centrifugation.

2.7.2 Partial specific volumes of selected amino acids

The amino acids are taken from Cohn and Edsall (1943, Table 3, Chapter 6). It is a tribute to the careful measurements made by these workers that, despite the improvements made in instrumental techniques over the years, these values remain unaltered. For example, the apparent molal volume (from which these values were calculated) of α-alanine was given as 60.6 $cm^3 mol^{-1}$ by Cohn and Edsall (1943, p. 159) and 60.47 $cm^3 mol^{-1}$ by Ahluwalia *et al.* (1977). The values for the specific volumes were calculated by Cohn and Edsall from the apparent molal volumes by assuming Traube's rule could be used to correct for the changes produced in the condensation to produce a polymer (removal of one molecule of H_2O per residue and the concomitant elimination of the charges on the residues). Thus, the apparent molal volume must be reduced by 7.4 $cm^3 mol^{-1}$ before converting to an apparent partial specific volume. These values can be used to calculate the apparent partial specific volume of a protein; provided that the amino acid composition is known completely, then the proportional contributions can be calculated. It seems that the result will be about 2% lower than the true value of proteins in simple solvents (see Table 2.3).

Table 2.3

Amino acid	Partial specific volume
Glycine	0.64
Alanine	0.74
Serine	0.63
Threonine	0.70
Valine	0.86
Leucine and Isoleucine	0.90
Proline	0.76
Phenylalanine	0.77
Methionine	0.75
Cystine	0.61
Tryptophan	0.74
Tyrosine	0.71
Histidine	0.67
Arginine	0.70
Lysine	0.82
Aspartic acid	0.60
Glutamic acid	0.66
Glutamine	0.67

CHAPTER 3

From friction to flexing

The dynamic motions of macromolecules are of continuous interest to biochemists, and often these motions are interpreted through models which infer that the chains and groupings are solid units. This approach gives a familiar touch to macromolecules which makes it possible to discuss activities using common parlance. The coordinates for these models usually stem from X-ray diffraction measurements, although electron micrographs provide equally useful constructions for large particles (see reviews by Crowther and Klug, 1975; Amos, 1974). Transposing these constructions onto the properties of unconstrained molecules in solution may raise doubts that the coordinates remain invariant with conditions. The basic difference between crystals and dehydrated molecules on the grid of an electron microscope compared with those in solution is that only in solution can free translational movement of the molecule occur. Furthermore, although internal flexing of the chain relative to its centre of mass could occur in all states, in practice it occurs more freely in solution than in a crystal. These arguments concerning states of a molecule under different conditions become more persuasive when considering coordinates from electron microscopy, since here the particle is examined in a dehydrated state. Distortions caused by changing the environment are variable; nevertheless, for particles such as a virus, evidence shows that these large particles can form surprisingly rigid shapes which resist distortions upon removal of the solvent. In this chapter are discussed the theoretical and practical attempts to measure detailed shapes of macromolecules in solution as well as defining various intermolecular interactions in more detail.

Before entering into the forest of details, it is useful to explore how the detailed coordinates of a macromolecule are viewed when the molecule is in solution. Static models constructed from X-ray diffraction studies represent temporally averaged positions of atoms. However, it is known that each atom, group, and chain are not stationary in space but are continually oscillating about a mean position with relaxation times normally less than 10^{-8} s. The energy for movements of these units is derived from electronic and thermal sources, but, as far as the X-ray diffraction records are concerned, their origin is not important because the diffractions are collected from averaged positions. From the reflections, the groupings are assigned positions in space and from these positions emerges the overall shape of the macromolecule. Thus, when discussing shapes of macromolecules, the picture of a static structure emerges, but this occurs by expediency rather than absolute reality; even in a crystal, the molecules are vibrating about mean positions. These views are important when considering some of the theoretical and practical aspects of macromolecules that are discussed in this chapter.

3.1 Defining shapes

The intrinsic total energy of a gas molecule can be divided between the harmonic motions of electrons in the atoms, rotation, vibration, and stretching of the covalent bonds, and translational movement of the whole molecule. Chemistry is concerned largely with interactions between the outer (valency) shells of neighbouring atoms as well as the rotational and vibrational motions of atoms in the molecule. Then, through the resulting wave equations can be calculated the energies of the gas molecule. With large polymers in solution, this approach is less tractable because the solvent molecules restrict movements. Even though it is possible to calculate covalent interactions for some simple polymers, the introduction of non-covalent interactions (inter-molecular forces) makes the problem more complex, and discussions usually end up employing statistical averages. Among these forces can be listed the inter- and intrachain hydrogen bonding as well as coulombic and van der Waals' forces (see Chapter 1 for the relative energies). The structure formed through the covalent bonds is the primary matrix upon which the weaker forces act to produce the tertiary and quanternary structures of macromolecules. All levels of atomic interaction are biologically important, but often it is the stable folded chain that produces the specificity of the macromolecule. The result of the folding is a model containing topological details such as clefts, helices, pleated sheets, etc., which suggest a solid unit rather than one consisting of interacting alternating electric fields. Nevertheless, it is these solid models which one uses when discussing the shape of macromolecules.

Rarely can one single molecule be studied experimentally, although results from electron microscopy for large assemblies come close to this ideal. Normally, results are obtained from solutions containing large numbers of similar molecules to give a result which is averaged overall. Detailed data on the coordinates of the molecules cannot be calculated from such a system in solution because all the molecules are given freedom of translational as well as vibrational movements. To study the shape of macromolecules in this system means either that the solid shape must be neglected in favour of descriptions of detailed interactions between neighbouring groups of each molecule (measured using light and magnetic resonance spectroscopy) or that the details are ignored, making it possible to consider the average molecule as a solid body in a continuum (the solvent). Only the latter case can be studied hydrodynamically. Naturally, results from crystallographic, spectroscopic, and hydrodynamic experiments must eventually converge to produce a consistent picture which describes the activity of the molecule in solution, but this ideal is rarely achieved with natural macromolecules.

3.1.1 Which technique is best?

The experimental times required to record crystallographic and hydrodynamic data are long compared with even the slowest molecular motions. The result is an average, and if, for example, two shapes are equally stable, then these two techniques would give average dimensions weighted by the proportions of molecules present in one shape compared with that in the second. Spectroscopic methods (in particular, magnetic resonance and fluorescence spectroscopy) provide detailed information about the two conformations because these methods measure electromagnetic interactions between the groups and the exploratory fields. So the important question is: Why continue to make measurements by crystallographic and hydrodynamic methods when spectroscopy can provide a more complete answer? At present, the problem raised by spectroscopy is how to interpret the detailed results that emerge from the experiments made on macromolecules containing many different functional groups. Usually, spectroscopy is better employed in answering specific questions about interactions between defined groups rather than trying to construct the whole macromolecular structure from perturbed resonances. Extensive structural interpretations may become routinely possible in the future but at present the application is limited in range so that details obtained from spectroscopy can only be filled into the overall solid models produced by crystallography. Purists might argue also that, because spectroscopic signals are often weak and it is necessary to work with concentrated solutions, this leads to abnormal intermolecular interactions. Extending this argument would eventually destroy the validity of data from crystallography since here it is the intermolecular interactions that give stability to the crystal. Clearly the logic is incorrect because experience shows that crystallographic data are important to discussions of biological activity of macromolecules. It must be remembered that the physiological concentration of many biological macromolecules is high, so arguments about concentration are only important when interpreting results from solutions and cannot be used to validate procedures. Hence, we have, on the one hand, spectroscopic measurements, which provide details of electronic interactions, and, on the other, crystallography, which gives detailed structural information. Hydrodynamics is complementary to both since it cannot yield detailed structural information nor specific group interactions but will provide data on the bulk shape of the solvated molecule as well as averaged molecular interactions.

3.1.2 What are hydrodynamic shapes?

If we were like Alice and could reduce our size so that we could see molecules directly, then this question would be even more complex than imagined. Everything would be moving and no molecule would retain a given shape for long periods. Fortunately, we are unable to emulate Alice, so we can use the parlance of the ideal world where shapes are readily defined through cartesian coordinates. If solid objects are transferred to dense fluids, they either sink or float depending on their densities relative to the solvent. The path taken when they sink will in part be determined by their shape; for example, a ball-bearing sinks in water by following a relatively straight line, whereas, when the same mass is formed into a thin rod or disc, it sinks more slowly while slipping from side to side. Basically, this view is the one used to describe the motion of macromolecules in a fluid, except that now the masses are usually insufficient to provide enough kinetic energy that they sink under the influence of the Earth's gravity. It follows from this view that the molecules must be doing work on the solvent during their movements and this work is converted to heat. It is from this and other sources of heat that the solvent molecules get their translational and rotational energy. These kinetic motions of the solvent molecules produce Brownian motion of the macromolecule to produce thoroughly mixed solutions; this translational motion can be readily overcome by increasing the gravitational field, as in the centrifuge (Chapter 2). Brownian motion can be regarded as a frictional energy which opposes movement of the macromolecule in a straight line. The difference in friction between spheres and rods reflects the increased surface area of the rod over that of the sphere having the same mass. The larger the surface area, the greater the probability that thermally energized solvent molecules will collide with the macromolecule. This impact will produce a couple which rotates the rod only if the body is asymmetrical, but translational movement of all shapes of bodies are affected by impacts with solvent molecules.

Although this oversimplified picture is useful in describing the motion of solid objects in a fluid, it completely ignores effects that occur at finite concentrations of the solute, since here there is a finite probability that two solute molecules will collide. The simple model cannot be readily extended to this situation because it assumes a discrete interface between the solid object and that of the continuous fluid. In the simple model, interaction between bodies can occur only when they are physically in contact. Everyday experiences with Venturi interactions show that there is a continuously changing velocity as one moves from one moving body to another even though the two bodies are not in physical contact. A simple example is a moving vehicle passing a stationary person; the person feels an increase in the pressure of the air as the vehicle approaches and a suction as it passes. Effects similar to this occur when molecules pass one another, but at this level one is concerned with molecular energies and coordinates. As discussed in Chapter 1, a solute molecule modifies the arrangement of water molecules at distances that extend beyond the individual water molecules adjacent to the surface of the solute molecule. Thus, for a moving macromolecule, there will be a gradual change of velocity from that of the moving molecule into that of the 'stationary' solvent (it is only stationary relative to the macromolecule; really all the molecules will be moving relative to the experimental frame of reference, say a breaker). This gradient in velocity is generally incorporated into the hydrodynamic shapes of solvated molecules; each molecule carries layers of solvent with it when it moves. The solvent molecules in this layer may be in complete equilibrium with the bulk of the solvent, but this does not affect the temporally averaged solvation.

A further point to consider when discussing hydrodynamic shapes of macromolecules is the effect that averaging has on the result (see Section 3.1.1). Although the individual molecules may be flexing, the hydrodynamic results will be an average for all the shapes in solution. This average over all the tumbling molecules will be recorded as a regular void in the continuous solvent that has an effective radius (if spherical) which is larger for an asymmetrical molecule than expected for the same number of spherical molecules of the same mass. Thus, descriptions of these averaged hydrodynamic shapes must involve mathematical models which relate molecular shapes to an oblate or prolate ellipsoid whose radii reflect the dimensions of the original molecule. Although coordinates for these models can be calculated from X-ray crystallography (as discussed later), it is difficult to reverse the procedure, so experimental hydrodynamic measurements cannot be used to calculate detailed shapes of a macromolecule. The result is always a smoothed ellipsoid.

3.2 Frictional coefficients

3.2.1 Known geometrical shapes

Translational movement

The qualitative view which relates friction to shape makes it possible to formulate mathematical models which contain measurable variables. The conversion from theory to practice is invariably in the direction of calculating Newtonian frictional coefficients to see if they correspond with experimental results; if they do not, then the geometrical shape must be changed and the calculation repeated. Possibly the best-known equation for calculating a frictional coefficient is for a sphere and was derived by Stokes in the middle of the nineteenth century

$$f_0 = 6\pi\eta a \qquad 3.1$$

Here, the frictional coefficient (f_0, having units kg s^{-1}) is related to the viscosity of the fluid (η) and the radius of the sphere (a) for a solid which is impenetrable to the solvent. This definition of f_0 is derived from Newton's laws for forces on a body, where the forces on a moving object equal the frictional coefficient multiplied by its velocity. Equation 3.1 has formed a cornerstone in all discussions of hydrodynamic shapes, and from it emerged Perrin's (1934) well known relationships for the translational coefficient (f_T) of an ellipsoid having a semi-major axis l and a semi-minor axis d, with $p = l/d$, i.e.

$$f_T = 6\pi\eta a\left\{(1-p^2)^{1/2}\Big/\ln\left[1+\left(\frac{(1-p^2)^{1/2}}{p}\right)\right]\right\} \qquad 3.2$$

Similar equations have been derived for cylinders (Broersma, 1960; Burgers 1938; Bloomfield *et al.*, 1967a, b) which include allowances for lack of streamline flow around the ends of the cylinder. These later models have their origins in the fundamental work of Kirkwood (1949, 1954) on irreversible processes in solutions of macromolecules. These studies show that, since the equations differ with shape of macromolecule, the conversion of molecular coordinates to frictional coefficient can only be achieved if the shape is known.

Rotational movement

Equation 3.2 applies to translational movement of the whole macromolecule within the fluid, but, of course, these molecules will be rotating about their centre of mass, this being caused by a turning moment induced by random hits from the solvent molecules. Normally, it is assumed that the rotation is random and decoupled from the translation (small velocities of translation in a Newtonian fluid). Naturally, the rotation can only be defined for asymmetric macromolecules and no rotational friction is measurable for an isotropic molecule (spherical). Thus, for a rod, two rotational frictional coefficients can be defined (f_l and f_d having dimensions length^2 mass/time) (Koenig, 1975)

$$f_l = 4f_1(1-p^2)/3(2-p^2S) \qquad 3.3a$$

$$f_d = 4f_1(1-p^4)/3p^2[aS(2-p^2)-2] \qquad 3.3b$$

$$f_1 = 8\pi a^3 p^2 \eta \qquad 3.3c$$

$$S = (2/a)(1-p^2)^{-1/2} \ln\{[1+(1-p^2)^{1/2}]/p\} \qquad 3.3d$$

In eq. 3.3, S is defined for a prolate ellipsoid (eq. 3.3d), $p < 1$; for an oblate ellipsoid, $1-p^2$ becomes p^2-1.

Studies of both translational and rotational frictional coefficients are simplified if the measured coefficients are related to the equivalent frictional coefficient of a sphere equal in volume to the ellipsoid, cylinder, etc. (a in eq. 3.1 would be replaced by a_e) (Tanford, 1961, p. 357). Thus, from eq. 3.2 the ratio would be $f_T/6\pi\eta a_e$ and the parts bracketed $\{\square\}$ would represent the asymmetrical contribution. Similarly, for eq. 3.3, f_1 is the hypothetical rotational frictional coefficient for a sphere.

3.2.2 Frictional coefficients for irregular geometrical shapes

Although it may be expediency that forces us to fit a macromolecule to one of the regular shapes, often this results in loss of information about the real shape of the macromolecule in solution. X-ray crystallographic structures show that, even if symmetry can be defined for a macromolecule, the peripheral boundary may not correspond exactly to a regular shape. Including irregular projections into an analytical equation is generally difficult, forcing workers to employ numerical procedures in order to include these irregularities. Among these methods, those by Adelman and Deutch (1975) and Bloomfield and co-workers (Bloomfield, 1966; Bloomfield *et al.*, 1967a; Bloomfield and De La Torre, 1977) are the best known. The method described by the latter is novel in that they represent the macromolecule by a rigid assembly of spheres, and this was successfully applied by Charlwood and Utsumi (1969) using earlier versions of the model. These workers all assume that the macromolecule can be represented as a thin shell containing n spherical subunits (radius r_e), each of which can be constructed from m smaller spheres of radius r. For this system, the frictional coefficient can be expressed by

$$f = \frac{6\pi\eta(\Sigma_{l=1}^{n} r_l^2)^2}{\Sigma_{l=1}^{n} r_l^2 + \Sigma_{l=1}^{n} \Sigma_{s=1}^{n} r_l^2 r_s^2 \langle R_{ls}^{-1} \rangle} \qquad 3.4$$

where $\langle R_{l,s} \rangle$ is the average distance between the l and s subunits (the sums are required only when $l \neq s$). Using eq. 3.4 and dimensions from electron micrographs for T2 bacteriophage, Bloomfield *et al.* (1967b) were able to calculate the maximal frictional radius to be 59 nm for the 'fast' form of the virus, and this value compared well with the experimental value of between 61 and 64 nm ('fast' refers to the sedimentation rates). Although it may appear that theory and experiment are not too similar in this case, it should be remembered that T2 has a head which is large in diameter compared with the stem (ratio of radii of head to stem is about 6.5) and a tail having six thin fibres each about 130 nm long. Each morphological part of T2 was modelled from contiguous spheres so, although the fibres may not be rigid in solution, the model assumes that they are rigid. Within these limitations, the calculated coefficient comes convincingly near to the experimental values. The success of this method improves with increasing asymmetry and size of macromolecule. Other models employing beads or spheres have been described by McCammon *et al.* (1975) and Nakajama and Wada (1977).

These equations show that the values of the coefficient undergo the greatest change with asymmetries (d/l) ranging between 1 and 0.2; outside this range, the changes are small (this follows from the logarithmic function which includes $(1 - p^2)^{1/2}$ in eq. 3.2). Frictional coefficients are useful parameters for testing whether shapes calculated from molecular ordinates (as for T2) can be recognized in solution. A further important application is in the detection and measurement of flexing of macromolecules, but before this application can be discussed in detail it is necessary to show how the frictional coefficient is related to flexing of a molecule.

3.2.3 Flexing of macromolecules

Flexing is an experimentally averaged parameter which describes molecular motions about a centre of mass. For example, flexing would be one possibility in explaining the finding that the experimental frictional coefficient was different from that calculated from molecular coordinates when a rigid macromolecule was used as the model (e.g. tail fibres of T2, Section 3.2.2). Naturally, the experimental result could only be collected by using a technique which has a long measurement time relative to the relaxation time of the macromolecule. (The definition of relaxation times comes from the original definition of a frictional coefficient and is inversely proportional to this coefficient: the relationship includes mass of the macromolecule, large molecules having longer relaxation times than smaller molecules having the same d/l ratio). Thus, if conventional hydrodynamic methods such as mass diffusion or intensity light scattering were the only methods available for measuring flexing, then the concept would be only of theoretical interest. Inaccuracies in measured frictional coefficients could easily encompass the difference between experimental and calculated values for most perturbations. Before examining the more accurate experimental methods, a discerning scientist would wish to know why flexing is important to describing the activities of macromolecules in solution.

Relaxation times

Motions of groupings on a chain can be studied through infrared, microwave, and Raman spectroscopy; but these methods produce detailed results on microscopic motions, and here we are discussing flexing of large polyatomic parts of the macromolecule, that is the movement of one part of the macromolecule with respect to other parts. The relaxation times of these movements can be important in determining rate constants for macromolecular reactions (Pecht and Lancet, 1977). Instability in a bound substrate is induced largely through configurational strains in the liganded molecule produced through attachment of the substrate at multiple sites on the macromolecule. This means that attachment can only occur when the sites are spatially arranged within the distances determined by the type of substrate. Flexing of the chains will in part determine these spatial arrangements. The relaxation times of this flexing will be one of the many compounded factors which determine the time interval occurring between the substrate first becoming attached at one point and the generation of the strain energy caused by multiple attachments. The amount that the relaxation time contributes to the complete rate constant will depend on whether translational diffusion of both the enzyme and substrate are limiting factors in bringing the substrate into contact with the site. Where two macromolecules must meet, both having catalytic regions, then translational and rotational diffusion both limit the rates of reactions. Considerations like these mean that flexing is best described in terms of relaxation times.

Possibly the simplest point to start discussion on relaxation times of flexing is from Perrin's explicit expression for orientational and rotational relaxation times for an ellipsoid of revolution using the corrected equations given by Koenig (1975). Thus, for an ellipsoid having semi-axes, $l/d = p$, there are two relaxation times τ_1 and τ_2 given by

$$\tau_1/\tau_0 = f_2/f_0 \qquad 3.5a$$

$$\tau_2/\tau_0 = 2f_1 f_2/f_0(f_1 + f_2) \qquad 3.5b$$

$$\tau_0 = 4\pi\eta l d^2/kT = 4\pi\eta l^3 p^2/kT \qquad 3.5c$$

(see also eq. 3.3) where τ_0 is the relaxation time for an equivalent sphere and kT in eq. 3.5c introduces thermal energy into the equations ($k = 1.38\ldots \times 10^{-23}$ $\mathrm{JK^{-1}}$). Using eq. 3.5 it is possible to calculate relaxation times of macromolecules in any solvent (for water, $\eta = 10^{-3}$ $\mathrm{kg s^{-1} m^{-1}}$) using differing asymmetries.

Suppose that a macromolecule has two conformations, one having a semi-axis $l = 5$ nm and $l/d = 0.5$ and the second form produced by flexing of the chains where l is reduced to 4.5 nm ($l/d = 0.45$), then τ_0 for the expanded molecule would be 9.71×10^{-8}s, while that of the contracted version would be 8.57×10^{-8}s. The molecular coordinates used in this calculation are within the bounds expected for globular proteins and show that even for relatively large masses the time constant of the flexing is short. Furthermore, it shows that the difference between the two times is proportional to the percentage change in semi-axis l. With classical hydrodynamic methods, the relaxation times would be too short to measure and the results would be a weighted average. Since τ_0 varies with the volume (ld^2), the relaxation time increases in proportion to the mass. For example, the relaxation time for the rotation of tobacco mosaic virus (approximate dimensions 300 nm $\times$ 15 nm, a molecule having a relative mass of about 4×10^7 dalton) becomes about 2×10^{-5} s, bringing it within the range of some classical techniques.

Values of relaxation times

Many examples exist which show that for natural macromolecules the relaxation rates are within this calculated order of magnitude. Craig *et al.* (1971) found that the rate constants for adjacent base-pairing of nucleotides along a double-stranded chain is about 10^{-7} s, and this motion is a major limit for the unwinding of a double helix; this rate would be consistent with a limiting step being the rotational displacement of interacting groups. Zimm and Bragg (1959) show that, for helix–coil transitions of polyamino acids, the relaxation times are between 10^{-5} and 10^{-6} s. These changes involve movement of large masses and are orders of magnitude slower than times recorded for flexing of small molecules like butenene where the rate constants for the *trans–gauche* transitions are about 10^{10} s^{-1} (Morawetz, 1972). This last example of a small molecular transition provides a useful example for relating relaxation times to energies, since here the enthalpy of the *trans–gauche* transition is about 16 kJ mol^{-1}; longer times infer larger energy barriers. Conformational transitions over energy barriers greater than 42 kJ mol^{-1} can be followed by nuclear magnetic resonance spectroscopy where the flexing causes broadening of bands (relaxation times of 10^{-3} to 10^{-1} s).

Flexible macromolecules have relaxation times for internal motion of chains which span the limits given above, and measurements have been made on synthetic polymers which match those predicted theoretically. King and Treadway (1976) used quasi-elastic light scattering (Section 3.4.2) to examine the flexing of polystyrene under theta conditions (that is, dissolved in a solvent of cyclohexane at 35°C where concentration dependence is negligible) and non-theta conditions (solvent of butan-2-one at 25°C). This polymer can be regarded as a random coil having no preferred conformation other than those determined by bonding restrictions. Pecora (1968) has calculated the broadening of the Rayleigh linewidth (see Section 3.4.2) caused by flexing of random-coil molecules. This model contains contributions from centre-of-mass motion (translational diffusion) and from internal modes of motion. King and Treadway (1976) extended this model to include polydispersity of their samples of polystyrene, which in itself produces broadening of the Rayleigh line, but this has relaxation times determined by the masses of the macromolecules. A typical relaxation time for the internal motion of this molecule was found by their method to be 135 ps ($M_w = 5.4 \times 10^6$) which compared favourably with the theoretical prediction of between 110 and 135 ps. Whether the model used for these calculations can be applied to study flexing of native proteins is doubtful, since most of these macromolecules are not random coils. It may have applications in studies on proteins denatured in sodium dodecylsulphate and heat–denatured nucleic acids, but these studies have not been reported in detail. These values are useful in that they emphasize the long timescale found for flexing of carbon chains compared with transitions such as helix–coil (10^{-6} s, Zimm and Bragg, 1959). It would be expected that similar magnitudes would be found for vibrations of polyatomic groups in proteins.

The values found for this type of macromolecular flexing show that the relaxation times are very slow relative to the time constants describing the breaking of covalent bonds, and only about one-tenth less than the relaxation time for mass movements associated with translational diffusion. Hence, flexing may form a significant part of the total rate constant of enzyme reactions but its contribution must be less than mass diffusion; it follows that, if flexing is damped by the presence of neighbouring macromolecules, the result would be a reduction in rates. Whether this inhibition could be cumulative and form a molecular on–off switch for the enzyme is open to debate, but the effect cannot be ignored.

Harmonic oscillator model

Although experimental results show that relaxation times are related to the masses of the moving parts, it is not apparent how the mass and damping are incorporated into theoretical models for relaxation times. It is obvious from eq. 3.5c that τ_0 is proportional to a^3, the volume of the molecule, and this is functionally related to mass through the density. Extending this proportionality further means taking a new direction, and possibly the simplest approach is to assume that the motions of the chains can be described by a damped harmonic oscillator obeying Newton's equation

$$F = -Sx - Z\,\mathrm{d}x/\mathrm{d}t \qquad 3.6$$

where F is the force ($= m\mathrm{d}^2x/\mathrm{d}t^2$), S is the stiffness of the chain, Z is the damping or frictional coefficient, and x is the linear displacement. The effect of damping is to force the amplitudes of the oscillations to decay exponentially and follow a first-order rate equation for the amplitudes (A) of a molecule of mass m, thus

$$A = A_0 \exp(-Zt/2m)$$

From this can be defined the relaxation time

$$\tau = 2m/Z$$

The value of Z will be determined by the viscosity of the medium as well as by bond stiffness but, if the process is restricted to first-order processes, then Z is equivalent to the rate constant for the process. Using an exponential relationship to relate amplitude with Z, t and m means that it is possible to obtain three solutions to eq. 3.6 each having different boundary conditions (Pain, 1976, p. 29). These solutions show that, as the damping increases, the time to decay to 1/e of the original decreases, and the force required to maintain the amplitude of oscillation must be increased (extra work is done by the solvent molecules). It is difficult to give numerical values for Z or d^2x/dt^2, so even order-of-magnitude calculations cannot be made from eq. 3.6. Hence, the harmonic oscillator model only provides an insight into the effect that mass has on the magnitudes of the energies involved: a more advanced quantum-mechanical treatment is necessary to relate these parameters exactly. It is possible, however, to provide an example of the order of energy required to maintain the oscillations using the relaxation times calculated for helix–coil transitions (Zimm and Bragg, 1959) for a molecule of relative mass 40 000 dalton, provided this time is equated with the oscillatory frequency (v). Restricting the oscillation to a few ångströms means that the energy required for this transition is approximately equal to the kinetic energy of water molecules at 300°C (the kinetic energy of a simple harmonic oscillator averaged over several periods is $\frac{1}{2}m\omega^2 A^2$ where $\omega = 2\pi v$, the angular frequency, or $\omega = (S/m)^{1/2}$). Thus, increasing the oscillatory mass means that more energy must be absorbed from the thermally active water molecules in order to produce the equivalent flexing and, if the system were isolated (that is, an adiabatic system), it would cool. In most cases, this does not occur because the system absorbs heat from the surroundings.

3.3 Experimental parameters

3.3.1 Diffusion and friction

Theoretical discussions of the shape of a macromolecule through frictional coefficients give only limited hints on the measurement of this parameter. Two important facts emerge from the discussions: (*a*) the motions of the macromolecule must be random and (*b*) these motions are moderated or even induced by interaction with thermally active solvent molecules. These movements are observed with large particles as Brownian motion, and this motion forms the source of theories of diffusion. Diffusion has the characteristic property that the mean displacement of the particle ($\bar{x}$) is proportional to the square root of time: the square root of the diffusion constant D is the important part of the proportionality constant in this relationship, i.e.

$$\bar{x} = 2(Dt)^{1/2} \qquad 3.7$$

(eq. 3.7 has classical importance to physical sciences since it provided one of the first routes for estimating Avogadro's number). For many purposes, eq. 3.7 is not useful and it is often used (although not operational) to define diffusion through osmotic pressure (see Section 3.6), when it is found that D equals the osmotic gradient divided by the frictional coefficient of the molecule. From this definiton comes the relationship between rotational and translational diffusion constants (D_i) and frictional coefficients,

$$D_i = kT/f_i \tag{3.8}$$

Effects of solvation

For translational diffusion constants ($D_i = D_T$) (eq. 3.8), f_T must be calculated from experimental measurements that have been extrapolated to infinite dilution. Even then, this value is not a true measure of asymmetry of the solvated molecule. The layers of solvent surrounding the molecule are undefinable but are included in the experimental value of diffusion. The extent to which solvation influences the amount of friction remains to be detected, so the experimental frictional coefficient f_e does not equal that calculated from eq. 3.2 or 3.4. The influence of solvation can be accommodated by factorizing the translational frictional ratio (that is, f_T/f_0, where f_0 is the frictional coefficient of an equivalent sphere) in the following way

$$\frac{f_T}{f_0} = \frac{f_T}{f_e}\frac{f_e}{f_0} \tag{3.9}$$

where f_T/f_e is the part of the determined coefficient which is dependent on solvation, while f_e/f_0 is the asymmetric component (Oncley, 1941). It is normal to assume that

$$f_T/f_e = [1 + (w/\bar{V}\rho)]$$

where w is the weight of solvent of density ρ solvating one gram of pure solute having a partial specific volume $\bar{V}$. This simple relationship was derived by assuming that the molecule is spherical when $f \propto (M\bar{V})^{1/3}$. However, for most globular proteins, the lack of sphericity will not materially distort calculations.

Since sharp transitions between solvent and the surface of a solvated macromolecule do not exist, the amount of water that should be included in f_T/f_e is debatable. Various methods have been proposed for estimating the amount of hydration (see review by Kuntz and Kauzmann, 1974), but earlier studies used adsorption isotherms and calorimetry to determine the amount of water taken up. These methods gave values of about 0.3 g H_2O/g protein (see Table 1 in Kuntz and Kauzmann, 1974) for most proteins. These classical procedures are difficult to employ when only small quantities of macromolecules are available, and a method based on NMR measurements of spin–lattice relaxation times for water has been described (Kuntz, 1971; Kuntz *et al.*, 1969) which works with about 1 cm^3 of a 0.1% w/v solution. This NMR method utilizes the fact that, when water molecules are held rigidly in ice, the natural motions are restricted, so reducing the characteristic Larmor frequency below that of water in the liquid state. Since water molecules associated with the macromolecule have lower chemical activity than those in the bulk of the solution, their motions remain relatively undisturbed when the macromolecule is frozen to form ice. Hence, a frozen solution still gives a characteristic proton resonance spectrum caused by the solvated water and this spectrum can be used to estimate the weight of water (estimated from the area of the peak) associated with a given weight of macromolecule. This value is taken as the definition of the solvated layer. Results from this method give higher but reproducible hydration levels than from calorimetry. For example, bovine serum albumin gave 0.29 g H_2O/g protein by the adsorption isotherm method (Kuntz and Kauzmann, 1974), but gave 0.4 g H_2O/g protein from NMR (Kuntz, 1971): similar values have been found for several globular proteins (Kuntz, 1971) using NMR and are consistent with hydration values obtained by dielectric relaxation (Grant *et al.*, 1971). Tests with phosphoglycerate kinase suggest that asymmetry calculated from frictional ratios obtained after correction for solvation and estimated from NMR were reasonably consistent with ratios calculated from overall dimensions taken from X-ray crystallography but not when lower levels of hydration were assumed (Smith and Spragg, 1979). Further discussions in this chapter assume that these corrections for solvation have been made when calculating frictional coefficients for relating to symmetry.

3.3.2 Concentration dependence of diffusion coefficients

In deriving eq. 3.8, it was assumed that the solution was ideal and that diffusion was not affected by concentration of the macromolecule. Experiments show that, like all hydrodynamic and thermodynamic variables, the diffusion coefficient is not constant with changing concentration. Further, since the driving force in diffusion is derived from changes in the chemical potential of the solvent (hence, according to the Gibbs–Duhem relationship, Chapter 1, a change in the chemical potential of the solute), the concentration dependence is related to thermodynamic functions. Equation 2.23 gives the conventional expressions which relate D to $(\partial \ln y/\partial c)_{T,P,n_2,\ldots}$ (y being the activity coefficient), but for many purposes it is more convenient to express the derivatives in a form of a series containing virial coefficients (Chapter 1). This means that eq. 3.8 must be expanded to contain at least the second virial coefficient when working at finite concentrations. A second source of concentration dependence arises from the dependence of the frictional coefficient on concentration. These two sources can be incorporated into a working equation by factorizing the concentration dependence into a true virial coefficient (B_2) depending on the excluded volume of the macromolecule (Section 1.2.3) and a second coefficient which describes the concentration dependence of the frictional coefficient (B_2'). Combining these factorizations with eqs. 1.15 and 1.16 gives the accepted relationship for the concentration dependence of diffusion:

$$D = (\partial\Pi/\partial c)/f = \frac{kT}{f}\frac{(1 + B_2 c)}{(1 + B_2' c)} \qquad 3.10a$$

$$D = D^\circ[1 + (B_2 - B_2')c] \qquad 3.10b$$

Comparing eq. 3.10 with eq. 2.21 (which describes the concentration dependence of sedimentation) shows that, at dilute concentrations, B_2' can be equated with g (eq. 2.21). For hard spheres, B_2' measures the effective volume ($\bar{V}$) available for the molecule to move in, and Batchelor (1972) showed that, for a neutral molecule, $B_2' = 6.55 \times \bar{V}(\mathrm{gl}^{-1})$. Experimentally, it is found that the net concentration dependence of diffusion ($B_2 - B_2'$) is small for neutral or ionically shielded molecules (Figure 3.1, showing the results for bovine serum albumin at a pH below its isoelectric point at ionic strength 0.2 M; Raj and Flygare, 1974). When the shielding is reduced (Figure 3.1, ionic strength 0.01 M), the dependence shows a relatively large positive relationship ($\mathrm{d}D/\mathrm{d}c = 2 \times 10^{-13}\,\mathrm{m^5\,g^{-1}\,s^{-1}}$ for bovine serum albumin, Figure 3.1). At pH 7.4 and ionic strength 0.15 M, the positive dependence is retained at a reduced level ($\mathrm{d}D/\mathrm{d}c = 0.5 \times 10^{-13}\,\mathrm{m^5\,g^{-1}\,s^{-1}}$), but at its isoelectric point (pH 4.7) this dependence becomes negative ($\mathrm{d}D/\mathrm{d}c = -0.7 \times 10^{-13}\,\mathrm{m^5\,g^{-1}\,s^{-1}}$, Fair *et al.*, 1978). These profound changes in sign of the dependence are reflected by most proteins as the pH is adjusted on either side of their isoelectric points (Timasheff *et al.*, 1957), and they reflect the changing solute–solvent–solute interactions. If the solvent conditions are not affecting the shape of the molecules, then D° (diffusion coefficient at infinite dilution) remains constant, as in Figure 3.1 where D° for bovine serum albumin at all the ionic strengths was constant at $4.9 \times 10^{-11}\,\mathrm{m^2\,s^{-1}}$ at pH 3.65. When the pH was increased, D° also increased to give a value of about $5.9 \times 10^{-11}\,\mathrm{m^2\,s^{-1}}$ at pH 7.0, so reflecting the contraction

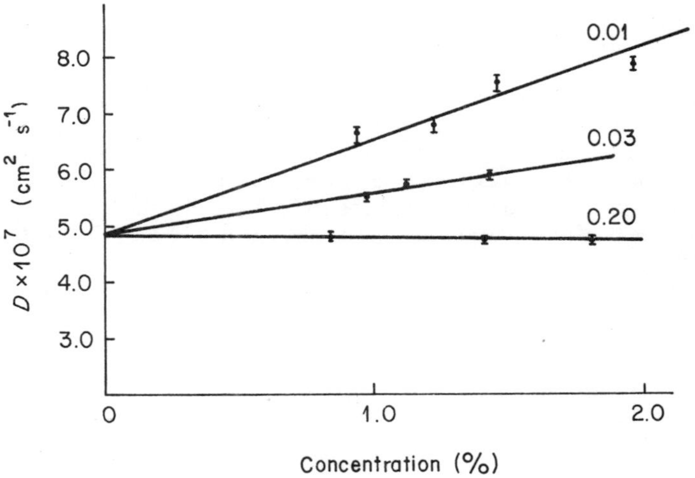

Figure 3.1 The dependence of the diffusion coefficient (D) on concentration (expressed as a volume fraction, Φ) for bovine serum albumin at pH 3.65 (20°C) and ionic strength 0.01, 0.03, and 0.2 M. (Taken from Raj and Flygare, 1974, Figure 6. Reprinted with permission from *Biochemistry* **13** 3336. Copyright by the American Chemical Society)

Table 3.1 Typical radii of bovine serum albumin calculated assuming that the protein was a hard sphere (r_{hs}) and taken from Raj and Flygare (1974, Table II). The results from Tanford *et al.* (1955) were obtained by viscometry

pH	Ionic strength (M)	$B_2 \times 10^3$ ($l\,g^{-1}$)	r_{hs}(nm) (Raj and Flygare)	r (nm) (Tanford *et al.*)
3.6	0.03	17.4	3.81	4.4
3.6	0.01	49.1	5.47	4.9
3.1	0.03	25.2	4.40	4.8
3.1	0.01	92.4	6.75	5.5

of the molecule at higher pH; as would be expected, these changes reflect those found for sedimentation coefficients (Section 2.4.1). Occasionally, one finds cases where the buffering ions affect the shape in a manner independent of pH, but unfortunately this is often ignored, as in the results of Fair *et al.* (1978) who used acetate buffer at low pH. These workers appear to have neglected the effect that diffusible molecules can have on shape in their analysis, and forced the results from differing buffer systems always to converge at the same D°. The results of Chiancone and Gilbert (1965, and shown in Figure 1.4) show that this can be dangerous procedure because the undissociated acetic acid penetrates the chains and expands the molecule, which in turn reduces the diffusion coefficient.

Evidence that B_2' can be equated with concentration dependence of sedimentation (eq. 2.22) is given by experiments such as those described by Raj and

Table 3.2 Average radii of bovine serum albumin at pH 4 (less than isoionic) calculated by a variety of methods and taken from Squire *et al.* (1968, Table V). The values were calculated using a constant axial ratio of 0.286. Here, V_e is the effective volume for an equivalent sphere, η is the viscosity of the solvent, $[\eta]$ is the intrinsic viscosity of albumin, ξ_{rot} is the rotational diffusion coefficient about the b axis, v is the Simha viscosity factor (value 4.17; see Tanford, 1961, p. 452) for an axial ratio of 0.286, V_{H_2O} is the volume of water, and V_p is the volume of protein

Equation		Radius (nm)	Hydration (V_{H_2O}/V_p)
$V_e = \dfrac{100\,M}{\mathcal{N} v}$	(viscosity)	2.86	0.21
$V_e = \dfrac{4\pi}{3}\left(\dfrac{f_e}{6\pi\eta}\right)^3$	(translational diffusion)	3.07	0.49
$V_e = \dfrac{\xi_{rot}}{6\eta}$	(rotational diffusion)	2.95[a] 3.08[b]	0.33 0.51
X-ray scattering (pH 5.1)		3.14	

[a] Obtained by dielectric dispersion (Moser *et al.*, 1966).
[b] Obtained from birefringence decay (Moser *et al.*, 1966).

Flygare (1974, from which Figure 3.1 is taken). These workers calculated B_2 from the experimentally observed concentration dependence of diffusion coefficients $(B_2 - B_2')$ by inserting a value for B_2' from the concentration dependence of sedimentation. Then from B_2 they calculated equivalent radii for the protein using equations which related B_2 to molecular coordinates (such as eq. 1.23). When these values were compared with radii calculated from viscosity (Tanford *et al.*, 1955), there was reasonable correspondence between the various estimates (Table 3.1). Unfortunately, close correspondence could not have been expected because the results included a variety of ionic strengths and the effective volume of the molecule increased with decreasing ionic strength, so changing the concentration dependence (Table 3.1). Comparisons of this nature give confidence in the view that differing hydrodynamic procedures can be related through overlapping parameters. Squire *et al.* (1968, shown in Table 3.2) provided a more comprehensive comparison of the methods, showing that the estimates of the effective volume of bovine serum albumin were similar when calculated from viscosity, dielectric dispersion, optical birefringence, and X-ray scattering. The results given in Table 3.2 also emphasize the sensitivity of these calculations to differing values assumed for the hydration of the protein. When these values are estimated from the hydrodynamic coefficients, they show wide variation, and one interpretation is that hydration is an accommodating coefficient which collects the experimental error rather than reflecting differing thicknesses of the water shell for different techniques.

Diffusion and osmosis

The molecular basis of the concentration dependence of diffusion which leads to B_2' being separated from B_2 means that the concentration dependence of the frictional coefficient can be separated from the thermodynamic non-ideality. This separation can be followed numerically by using the concentration dependence of osmotic pressure to calculate a series of diffusion coefficients corrected for virial effects (eq. 3.10a). These can then be compared with experimental values for diffusion and the differences discussed. This calculation is made possible by the work of Ross and Minton (1977) who analysed the earlier osmotic data of Adair (1928) for oxyhaemoglobin (discussed also in Chapters 1 and 2, Figure 2.1). Ross and Minton (1977) showed that it was necessary to estimate only two parameters in order to model the concentration dependence of Adair's data over a very wide range of concentrations (up to about 340 $\mathrm{g\,l^{-1}}$). These parameters were relative mass and the second virial coefficient B_2. This treatment assumes that this coefficient could be expanded through known relationships in order to accommodate results at high concentrations of hard spheres (Hirschfelder *et al.*, 1967). Several estimates of B_2 are available for oxyhaemoglobin, all agreeing reasonably well with the original estimate of Adair (1928) and giving a value of 3.181 $\times 10^{-3}\,\mathrm{l\,g^{-1}}$. Then, assuming that tetrameric oxyhaemoglobin could be represented by a hard sphere with no short-range interactions between the spheres (that is, non-ideality was caused by exclusion of solvent volume and no

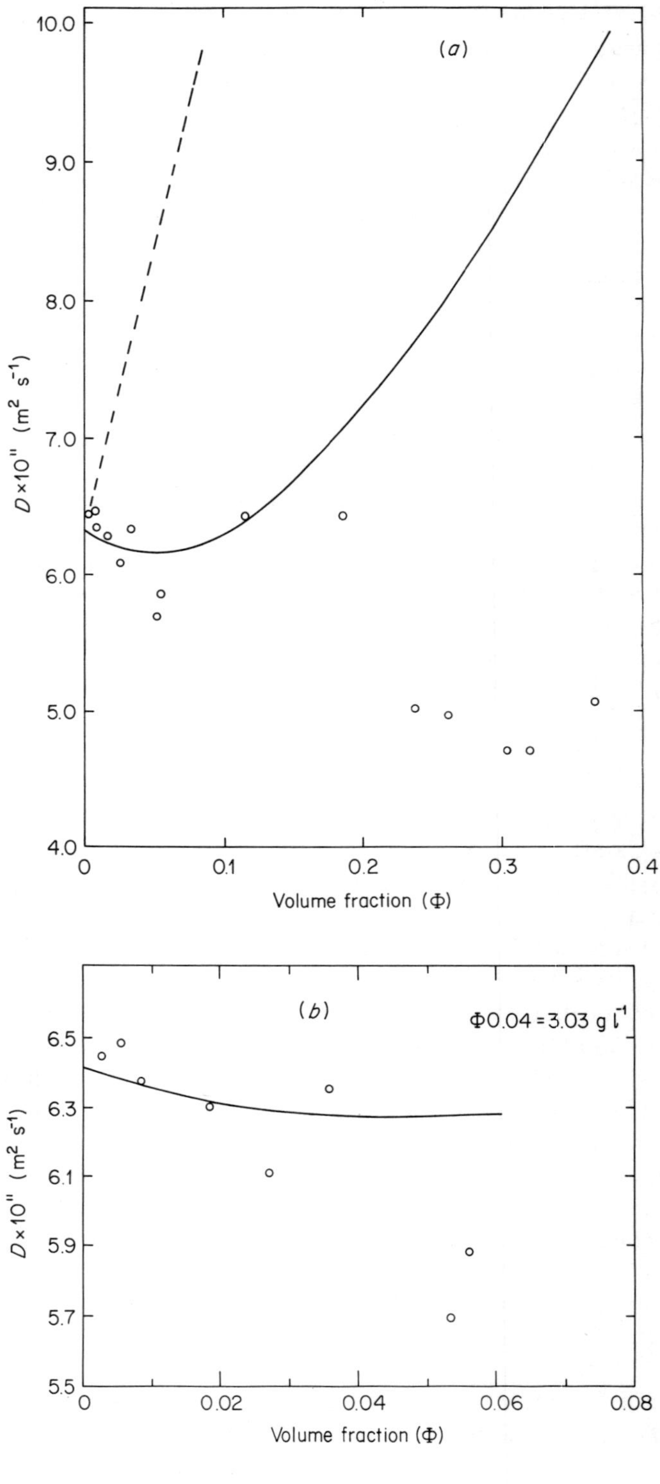

(a)
10.0
9.0
8.0
7.0
6.0
5.0
4.0
D×10'' (m² s⁻¹)
0
0.1
0.2
0.3
0.4
Volume fraction (Φ)
(b)
Φ0.04 = 3.03 g l⁻¹
6.5
6.3
6.1
5.9
5.7
5.5
D×10'' (m² s⁻¹)
0
0.02
0.04
0.06
0.08
Volume fraction (Φ)

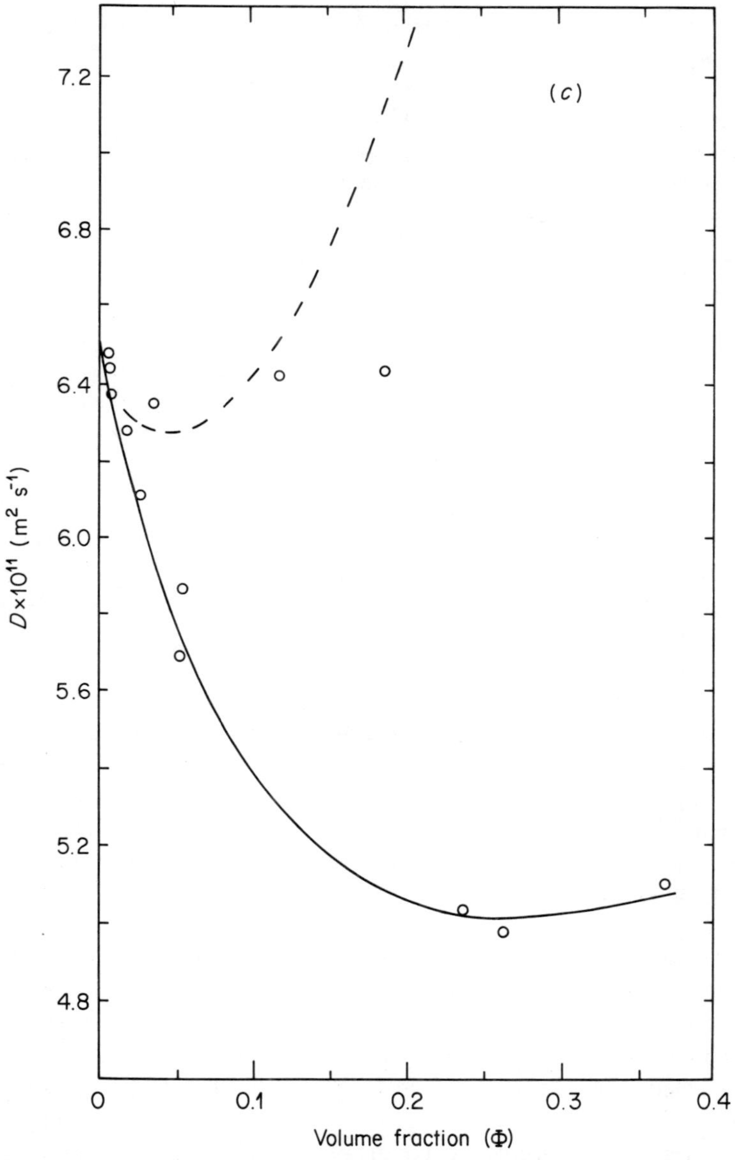

Figure 3.2 Relationship between experimental values of diffusion (0) obtained by Alpert and Banks (1976) for human haemoglobin and diffusion calculated using osmotic coefficients using eq. 3.12. The equation of Ross and Minton (1977) (eq. 3.11) was used to describe Adair's (1928) original osmotic data for sheep haemoglobin. The value for G (eq. 3.12) was -5.12 (Alpert and Banks, 1976). (*a*) Results for $B_2 = 3.85 \times 10^{-3}\,\mathrm{l\,g^{-1}}$. (*b*) Detail from (*a*) showing the fit over the range of volume fractions 0 to 0.06. (*c*) Curve for $B_2 = 2.5 \times 10^{-3}\,\mathrm{l\,g^{-1}}$. In (*a*) the dashed line is the calculated curve uncorrected for G (no frictional concentration dependence), while the simulation given in (*a*) is repeated in (*c*) for comparison

electrostatic effects were included by working at pH 7.0, the isoelectric point), then B_2 can be expanded into a series using the relationship

$$\Gamma_i = p_i \Gamma_2^i \quad (\text{for } i \geq 3)$$

where p_i are numerical coefficients calculated using the methods of Hirschfelder *et al.* (1967) and Ree and Hoover (1967). The relative contribution given by individual coefficients depended on the absolute concentration (c_2), but $n \leq 7$ (eq. 3.11). Thus, the full equation for relating osmotic pressure to concentration becomes

$$\Pi = \frac{RT}{M_2}\left(c_2 + \sum_{i=2}^{n} \Gamma_i c_2^i\right) \qquad 3.11a$$

$$\Pi/c_2 = \frac{RT}{M_2}\left(1 + \sum_{i=1}^{n} p_i (B_i c_2)^i\right) \qquad 3.11b$$

(Note that $p_1 = 1$, while the other coefficients are listed by Ross and Minton (1977) and Alpert and Banks (1976).) It is possible to calculate a value for B_2 for a hard sphere using eq. 1.23 ($B_2 = 4\bar{V}$) but this value did not correspond with volumes expected from X-ray measurements. Converting the experimental value for B_2 into a normalized coefficient by dividing by the relative mass (64 500 dalton) gives 5.9×10^{-8} lg^{-2} mol^{-1} and would correspond to a hard sphere of radius about 5 nm (X-ray crystallography shows that oxyhaemoglobin has an effective volume of 3.4 nm). This discrepancy remains unexplained, but if one accepts the experimental value for B_2 then it is possible to calculate diffusion coefficients from the data.

First, the diffusion coefficients were calculated from the osmotic gradient ($d\Pi/dc$) using eq. 3.11 and a value for the frictional coefficient calculated using eq. 3.1 for a hard sphere. These coefficients could then be compared with experimental values for diffusion measured by Alpert and Banks (1976) and covering a similar range of concentrations. This comparison will only be meaningful if the sources of diffusion are comparable, which means that they both have the same frames of reference. The frame of reference for osmotic pressure is the movement of the solute with respect to the solution, but in measurements of diffusion it is related to the coordinates of the apparatus. Keeping the same frames of reference is important, as can be seen from Figure 3.2 where the broken curve is drawn from $(d\Pi/dc)/f$ without making allowance for either the frame of reference or dependence of f on concentration. Even after changing the frame of reference, the calculated and measured results are still not in agreement (the change is made through allowing for densities of solute and solvent through the volume fraction Φ to give $1 - \Phi$, eq. 3.12 and given in Section 3.6.2). This first comparison shows that the simple relationship for relating diffusion to concentration given in eq. 2.23 is only applicable for situations where no nett flow of solutes occurs, as in sedimentation equilibrium.

A second correction must now be applied to include the concentration dependence of the frictional coefficient and this is a more complex procedure. The complete equation must include a full account of the solute–solvent–solute interactions and these can be collectively given by a further coefficient G, as in the equation given by Phillies (1975), to give the equation without virial corrections for diffusion, i.e.

$$D = \frac{(\partial \Pi/\partial c)_{T,P,n_2,\ldots}(1-\Phi)}{f(1-G\Phi)} \qquad 3.12$$

where Φ is the volume fraction calculated from concentration in weight per unit volume by multiplying by $4\pi\mathcal{N}r/3$ where r is equal to the hydrodynamic radius (including solvation). The complex coefficient is defined by Phillies (1975) through

$$G = [3/(4\pi r^3)] \int_V jD^{T}(r).j[\exp(-w(r)/kt) - 1] \qquad 3.13$$

where $D^T(r)$ is the cross-diffusion tensor related to the Oseen tensor, j is the unit vector in the direction of particle flow, and $w(r)$ is the nett energy of interaction between molecules in solution. This energy provides the important insight necessary for relating friction at the molecular level to the algebraic sum of attractive dispersion forces and repulsive electrostatic forces. Without these energies, G is inversely proportional to the macromolecular volume and could be calculated from molecular coordinates, but when these forces are included G cannot be readily calculated. Fortunately, a value for this integral can be estimated from the diffusion experiments of Alpert and Banks (1976) where they found it was equal to $-8.9 \times 10^{-19}\,cm^3$ for oxyhaemoglobin; the operative value for G was -5.8 under their conditions. When this value was used to calculate the diffusion coefficients from the osmotic data (eq. 3.12), the calculated values agreed with measurements only over a limited range of concentrations (Figure 3.2a and b). The virial corrections when combined with the hydrodynamic effects were still insufficient to compensate for the rapid increase in osmotic gradient at higher concentrations; the experimental values for diffusion continued to decrease linearly with concentration up to concentrations of about $180\,g\,l^{-1}$.

The theory seems exact, so the discrepancy can only be explained through the lack of interchangeability between the two sources of data. The major differences in experimental conditions were in the solvent and temperatures; Adair (1928) used phosphate buffer plus KCl at ionic strength approximately 0.3 M and pH 7.8 at 0°C for sheep haemoglobin, while Alpert and Banks (1976) used 0.1 M phosphate at pH 7 and at 20°C with human oxyhaemoglobin. This means either that the values for B_2 or G were too high or too low, respectively. It is impossible to choose between these alternatives with present data, but recalculating the diffusion using a lower value for B_2 ($2.5 \times 10^{-3}\,l\,g^{-1}$) gave a satisfactory fit over all concentrations (Figure 3.2c). This value for B_2 is equivalent to a normalized B_2 of $3.7 \times 10^{-8}\,l\,g^{-2}\,mol^{-2}$, which corresponds more closely with the radius for human oxyhaemoglobin estimated from X-ray crystallography than the value used originally by Ross and Minton (1977) but which described Adair's data so well. It seems likely that both the coefficients are affected by the differing conditions rather than the change of species, since B_2 is not significantly different between the species (summarized in Table 4 of Ross and Minton, 1977).

Intermolecular interactions

Two facts emerge from the attempt to relate osmotic data to diffusion: (*a*) apparently small changes in experimental conditions may lead to relatively large changes in B_2 and, hence, in the effective volume of the molecule, and (*b*) diffusion measurements covering a wide range of concentrations give information about the solute–solvent–solute interactions. Of particular value is the estimate for the concentration dependence of the frictional coefficient, which gave a value for the integral (eq. 3.13) of -8.2×10^{-19} cm^3 for oxyhaemoglobin (Albert and Banks, 1976). The square root of this (absolute) value gives an estimate of the radius of interaction, which in this case was 9.8 nm. This is about three times the true radius of the unsolvated molecule, indicating that the molecules move within an average sphere equal to three times the radius of the molecule. The interesting finding that the hard-sphere model described the properties of oxyhaemoglobin at its isoelectric point (Ross and Minton, 1977) gives encouragement for further work along these lines. However, whether the present calculations convince people that osmotic pressures and diffusion are manifestations of the same phenomenon depends on whether one believes the values for the estimated coefficients.

There is at least one other set of results which can be treated to give values of G, namely the measurements of the diffusion of myosin (Herbert and Carlson, 1971). This molecule can be hydrodynamically modelled as a cylinder having dimensions 130 nm × 5.5 nm bearing a sphere of diameter 29 nm at one end, so giving a total volume of about 7.3×10^{-18} cm^3 per molecule. The diffusional data of Herbert and Carlson (1971) are more complex than those for oxyhaemoglobin, since not only is there non-specific hydrodynamic effects but also there occurs reversible dimerization of myosin at the concentrations used for the analysis; this association was reported earlier by Godfrey and Harrington (1970a, b). Having to include two processes complicates the analysis because, in addition to estimating the concentration dependence of both the monomer and dimer, it was necessary to estimate an interaction coefficient. Nevertheless, the data of Herbert and Carlson (1971) were sufficiently precise that the three parameters could be estimated. The monomeric B_2 was found to be 0.321 $l\,g^{-1}$ while the measured concentration dependence was -5.6×10^{-3} $l\,g^{-1}$. Using these values made it possible to calculate B_2' (eq. 3.10b) for the monomer (0.327 $l\,g^{-1}$) and this gives a value for G (eq. 3.12) of -6.6 (the value used to convert concentration in $g\,l^{-1}$ to volume fraction was 1.15×10^{-3} $l\,g^{-1}$). From G was calculated the volume integral of -4.8×10^{-17} cm^3 which converts to a spherical radius of 36.0 nm and is about one-and-a-half times the radius of the equivalent ellipsoid for myosin (this is about 13 nm) calculated from B_2. The size and dimensions of myosin are intermediate between those of tobacco mosaic virus and oxyhaemoglobin, so linking the static arrays of TMV (discussed in Section 1.4.3) with the more dynamic arrays of concentrated myosin and oxyhaemoglobin. The forces between the solvent and macromolecule are similar in all cases, and it seems that the distance separating the macromolecules probably represets a minimum in the potential-energy profile. These curves are formed by balancing van der Waals' and other attractive forces against electrostatic forces (see Chapter 1 for a discussion of these forces). These forces are included in eq. 3.13 through $w(r)$.

Until recent times, most discussions of concentration dependence had been through the collective virial coefficients which carry simple connotations for neutral molecules but which are less obvious for charged macromolecules in an aqueous solvent. Tanford (1961) reviewed the calculation of the second virial coefficient for charged spheres and showed that for a sphere of valency z the relationship for B_2 becomes

$$2B_2M_2 = 8\bar{V}_2 + \frac{10^3\bar{V}_1z^2e^2}{2IM_2} \qquad 3.14$$

(this is an advanced version of eq. 1.25). Here $\bar{V}_1$ and $\bar{V}_2$ are the partial specific volumes of water and macromolecule, respectively, while I is the ionic strength of the solvent ($\frac{1}{2}\Sigma_i c_i z_i^2$). Figure 3.1 shows that, as the ionic strength is decreased the concentration dependence of diffusion increases for moderately charged macromolecules (bovine serum albumin). A more complex dependence is found for more highly charged macromolecules like t-RNA mixed with Mg^{2+}. As Figure 3.3 shows (taken from Olson *et al.*, 1976), with this macromolecule there

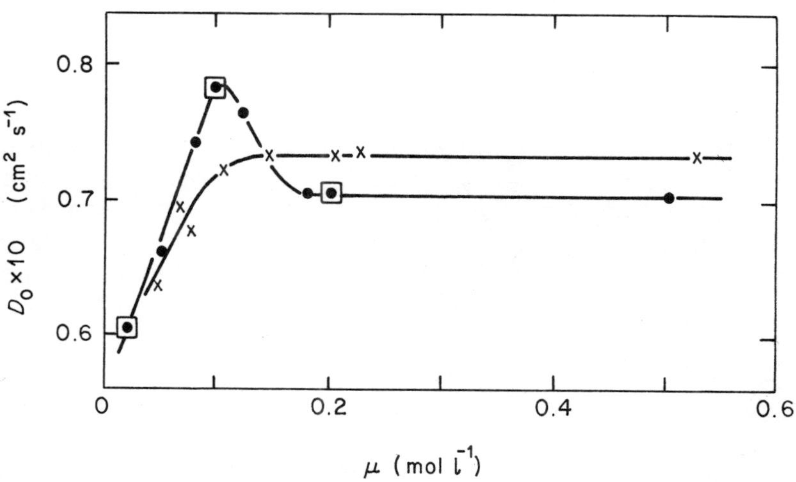

Figure 3.3 The effect of ionic strength (μ) and Mg^{2+} on the diffusion constant (D_0) of t-RNA: ●, unfractionated t-RNA from *E. coli* at 1 mM Mg^{2+}; ×, unfractionated t-RNA from *E. coli* in 10 mM Mg^{2+}; ⊡, measurements made with unfractionated t-RNA from *E. coli* and yeast (t-RNAS^{phe}, at 1 mM Mg^{2+}. (Taken from Olson *et al.*, 1976, Figure 3. Reproduced with permission from *J. Molec. Biol.* **102** 193. Copyright by Academic Press Inc. (London) Ltd)

occurs a maximum in the relationship between diffusion and ionic strength. This is interpreted as showing that increasing the ionic strength from 0.1 to 0.18 M was accompanied by a 11% decrease in the hydrodynamic radius when 10 mM Mg^{2+} was present; no similar change was recorded for 1 mM Mg^{2+}. This change from an expanded molecule to a more compact version is followed by a return to the expanded form if the ionic strength is increased further. Olson *et al.* (1976) conclude that this surprising effect must reflect a reorganization of the tertiary structure rather than a general shrinkage and expansion of the whole molecule. However, clear conclusions from this work are not possible because there is evidence of reversible dimerization of t-RNA in 10 mM Mg^{2+} as the concentration of nucleic acid is increased, and this association produces a negative second virial coefficient (Figure 3.4). However, the linear dependence of B_2M_2 on the reciprocal of the ionic strength (as predicted by eq. 3.14) is preserved over a wide range of ionic strengths at both concentrations of Mg^{2+}. Even larger charge effects are produced by some molecules, and Camerini-Otero *et al.* (1974) give an example of a virus (R17) which at pH 6.3 showed a positive concentration

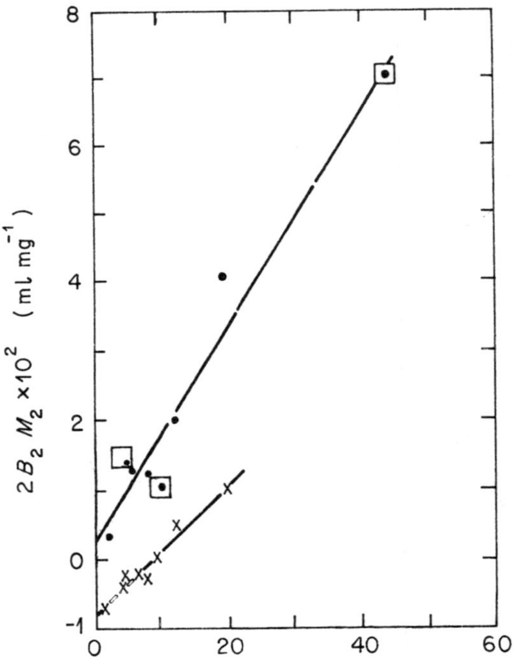

Figure 3.4 Effect of ionic strength (μ) on the second virial coefficient (B_2M_2) for t-RNA at two concentrations of Mg^{2+}: , 1 mM; ×, 10 mM; ⊡, as for Figure 3.3. (Taken from Olson *et al.*, 1976, Figure 2. Reproduced with permission from *J. Molec. Biol.* **102** 193. Copyright by Academic Press Inc. (London) Ltd)

dependence of the diffusion even in 1 M NaCl. Charge repulsions between macromolecules in 1M NaCl are rare unless the molecules are highly charged.

Theoretical studies of concentration dependence

Introducing an energy term ($w(r)$ eq. 3.13) into concentration dependence means that it is possible to use this dependence for studying solute–solvent–solute interactions; in fact, it may be one of the few methods available for measuring the energy of these interactions. The interpretation of this dependence has been through balancing the attractive forces ascribed to van der Waal's or dispersive forces, with coulombic repulsive forces. It may be thought from the discussions of viral coefficients (Section 1.2.3) and diffusion of neutral molecules or well shielded molecules (Figure 3.1, that is where $\kappa r \gg 1$, κ^{-1} being the Debye–Hückel screening length and r the radius of the macromolecular ions) that the attractive forces would be fully described through the model of excluded volume. Unfortunately, this is not supported experimentally because, as Fair *et al.* (1978) showed, bovine serum albumin at pH 7.4 in an ionic strength of 0.15 M still gave a positive concentration dependence for diffusion. Under these conditions, the molecule carries about 20 excess charges and should be reasonably well shielded, yet the dependence still shows repulsion between the molecules in solution. To model the dependence of diffusion on concentration, Anderson and Reed (1976) factorized the attractive potential between solute molecules into two parts, that from the hard sphere (excluded volume) and that from long-ranging forces formed from the sum of attractive dispersion forces (Section 1.4.3) and repulsive electrostatic interactions (Debye–Hückel, p. 26). The equations formed from this model fitted the results of Fair *et al.* (1978) reasonably well, although the calculations were made assuming that the contribution from the dispersive forces (eq. 1.40) was negligible. A similar approach was made by Goldstein and Zimm (1971) for modelling the concentration dependence of sedimentation (eq. 2.22) of polystyrene spheres by calculating the intrinsic viscosity of the macromolecule using a model based on the algebraic sum of these forces. In this case, the dispersion forces were included in the models, and this model gave a successful quantitative simulation of the concentration dependence of polystyrene using the data of Cheng and Schachman (1955). More exact formulations of the hard-sphere contributions are given by Batchelor (1972, 1976) and his value for the dependence differs from that found by Anderson and Reed (1976). Numerically, the two values for a hard sphere were similar for the two models (both less than 2, when concentration is expressed as a volume fraction) but differ in sign. Anderson and Reed (1976) found a small negative dependence of translational diffusion and concentration (indicating attraction between the spheres), while Batchelor's treatment (1976) gave a small positive dependence (indicating hydrodynamic repulsion). It may prove difficult to decide between these models experimentally because it is difficult to produce stable suspensions of uncharged spheres in water, and real tests can only be made for this solvent if Batchelor's model can be expanded to include solute–solvent and solute–solute interactions.

These models of concentration dependence show that solute–solute interactions for these (so-called) neutral spheres can extend as far as 1.5 nm out from the surface of the sphere. This represents a total diameter of about 10 water molecules, and hydrogen bonding between these molecules must represent some part of the electrostatic energy of these interactions.

Summarizing non-ideality and concentration dependence

Sufficient evidence has been presented to show that although the influence of a macromolecule on the solvent can extend out further than would have been expected intuitively from simple models of aqueous solutions (Chapter 1), the result for diffusion measurements is a nett concentration dependence which is usually negative and small. If either large negative or positive dependence is recorded, then either the macromolecular charge may be insufficiently shielded (positive dependence with repulsion) or the charge is favourably distributed to produce repulsion even in high salt concentrations; reversible association gives a larger-than-expected negative dependence. The positive dependence has been discussed for albumin (Figure 3.1) and t-RNA, while the negative dependence has been given for myosin with association: this last effect is discussed in general in Chapter 4, while two further relevant cases of the interpretation of association are given in detail here.

Macromolecular association and diffusion

Glutamate dehydrogenase catalyses the reversible oxidative deamination of glutamic acid through NAD^+ to give α-ketoglutamate. It was found that high-energy triphosphates (ATP, etc.) inhibited the reaction, while low-energy diphosphates activated the catalysis (Frieden, 1963); this action was interpreted as being an allosteric change in the protein. Further support for this interpretation was given when it was found that glutamate dehydrogenase reversibly associates in the presence of ADP but that association was reversed when ATP was added, suggesting that the aggregated molecule was the active unit (see Cohen and Benedek, 1976, for a concise review of these effects). However, subsequent work suggested that the aggregation was not important in determining the activity. Cohen and Benedek (1976) pursued this problem further by suggesting that the monomer could exist in two conformational states which were in reversible equilibrium (x and y), the equilibrium being determined by the presence of an affector, but both states could form polymers: only one of these states may be enzymically active. For this model, three chemical equilibrium coefficients are defined, K_x for polymerization of x, K_y for y, and K_0 for the interconversion between the states, which is defined as the ratio of concentrations of x and y in the mixture. The first two coefficients are independent of enzyme and affector concentrations, while K_0 is the only variable which is dependent on the affector concentration Several models are proposed but only the one described above fitted the diffusion record (Figure 3.5) and, since only low concentrations of

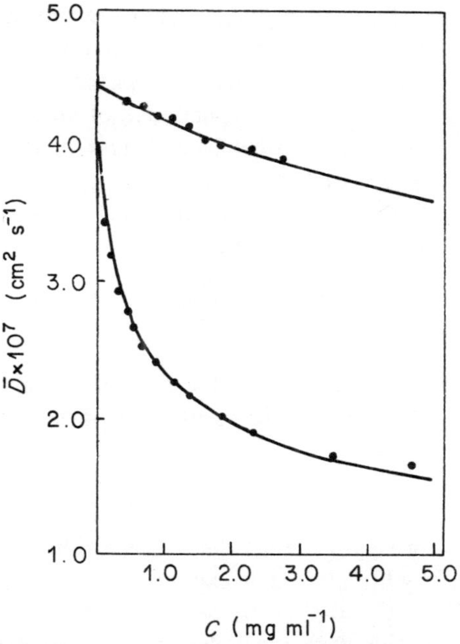

Figure 3.5 A plot of the weight-average diffusion coefficient ($\bar{D}$, corrected to standard conditions) as a function of the concentration of glutamic dehydrogenase (c). Data for the upper curve were measured at 25°C in solutions containing 20 mM GTP and 20 mM NADH in 0.2 M sodium phosphate buffer, pH 7.0. Data for the lower curve were obtained under identical conditions but in the absence of GTP and NADH. Continuous curves are theoretical fits using D(monomer) $= 4.44 \times 10^{-11}$ m^2s^{-1} and two equilibrium constants (see text). (Taken from Cohen and Benedek, 1976, Figure 1. Reproduced with permission from *J. Molec. Biol.* **108** 151. Copyright by Academic Press Inc. (London) Ltd)

protein were required for the study, the concentration dependence of the individual species could be ignored. The utilization of accurate primary measurements under conditions where the parameters can be taken as constants and where activity was equivalent to concentration, made it possible to fit a relatively complex model with confidence. Other hydrodynamic measurements show a larger concentration dependence than diffusion, so the picture would not have been so clear.

A similar and equally interesting application occurs in the kinetic study of the coupling of heads to tails to form non-infective fibreless T4 (Aksiyote-Benbast and Bloomfield, 1975). Here the important factor in the choice of diffusion was that the time taken to measure an average diffusion coefficient for the mixture of heads, tails, and fibreless T4 was small compared with the time taken for assembly. Again, the concentration dependence of all components was small, while it was assumed that the diffusion coefficient of the tail was sufficiently different from either the head or T4 that it could be separated experimentally. The technique used was homodyne light scattering (see Section 3.42) where the relative light intensities can be given by

$$I_t(\tau)/I_0 = A_{1t}^2 \exp(-2K^2 D_1 \tau) + 2A_{1t}A_{2t}\exp[-K^2(D_1 + D_2)\tau] + A_{2t}^2 \exp(-2K^2 D_2 \tau) \qquad 3.15$$

Here D_1 and D_2 are the diffusion coefficients of the head and fibreless T4, respectively; A_{1t} and A_{2t} are the relative scattering powers of head and T4 at time t, defined as

$$A_{it} = M_i c_{it}/(M_1 c_{1t} + M_2 c_{2t})$$

and K is the scattering vector. Thus, the average diffusion coefficient ($\bar{D}$) was estimated many times by fitting a single exponential, $\exp(-2K^2\bar{D}\tau)$, to each set of correlation data. These could then be assembled to give the result shown in Figure 3.6. It was found that with quasi-elastic light scattering it took only 10 s to measure a diffusion coefficient and this could be repeated every 30 to 40 s. The interval represents 'dead time' of the experiment and is insignificant compared with the assembly time. From the measurements, a second-order rate constant of $1.02 \times 10^7\ \mathrm{mol^{-1}\,s^{-1}}$ was calculated (20°C at pH 7.2).

These examples are taken from measurements of translational diffusion coefficients but few studies have been made on the concentration dependence of

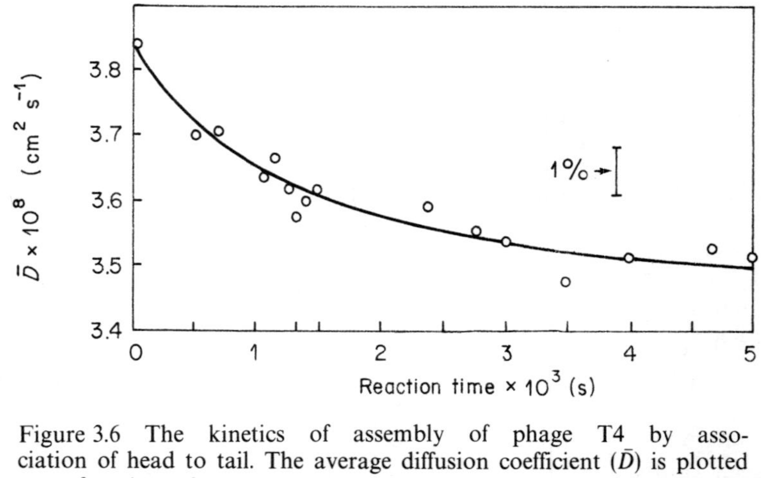

Figure 3.6 The kinetics of assembly of phage T4 by association of head to tail. The average diffusion coefficient ($\bar{D}$) is plotted as a function of time at 22°C. (Taken from Aksiyote–Benbast and Bloomfield, 1975, Figure 6. Reproduced with permission from *J. Molec. Biol.* **95** 335. Copyright by Academic Press Inc. (London) Ltd)

rotational diffusion (however, see Section 3.4.3 and Figure 3.11 for one study which suggests that this dependence may be large). From the translational studies, two contrasting approaches have been applied experimentally. In the one case, the concentration dependence is utilized for calculating overall shapes and hydrodynamic interactions. In the second case, the conditions are arranged to give small concentration dependence, making it possible to ignore the dependence and study other properties of the system; this would be difficult to achieve with other methods.

3.4 Estimating diffusional parameters

The experimental realization of theoretical predictions can often be difficult and measuring diffusion is no exception to this view. Experimental conditions must be arranged so that the movement of the molecule is random and driven by thermal forces; biased effects such as thermal density inversions or other sources of mixing must be avoided. Complete removal of all systematic errors has proved difficult when using classical methods, and measurements of diffusion of macromolecules covering a wide range of concentrations are tedious. These early measurements were made by studying the mass flow across boundaries when diffusion coefficients were estimated by fitting analytical or numerical solutions of Fick's laws of diffusion (see Section 3.4.1). The development of stable lasers has made it possible to move from boundary analysis to methods which record Brownian motions of macromolecules through the frequency shifts of scattered light relative to the incident frequency. With this method, it is possible to examine macromolecular solutions at true equilibrium and undisturbed by external forces except the laser light (which under some circumstances can produce a high electrical field) or requiring the formation of boundaries between two concentrations. These two contrasting methods are considered in Sections 3.4.1 and 3.4.2 and are collectively classified as normal methods for estimating translational diffusion coefficients; other less conventional procedures available for following relaxations which are best described through rotational diffusion are discussed in Section 3.4.3.

3.4.1 Diffusion by mass transfer

When a solvent is layered on a solution, the solute diffuses into the upper solvent. The concentration ($c_{x,t}$) at any position (x) within the column of liquid will vary with time, and its value reflects the continual diffusion of the solute from the high to low concentrations (Figure 3.7). Experimentally, the diffusion produces a sigmoid concentration curve, and records of this profile can be captured on film using a variety of optical techniques (described in Schachman (1959, p. 32) and Lloyd (1974); while Alexander and Johnson (1949, p. 239) give details of the complete experiment including the analysis of the records using classical techniques). After measuring the photographs, the ordinates are fitted to a suitable diffusion equation whose form varies with experimental conditions.

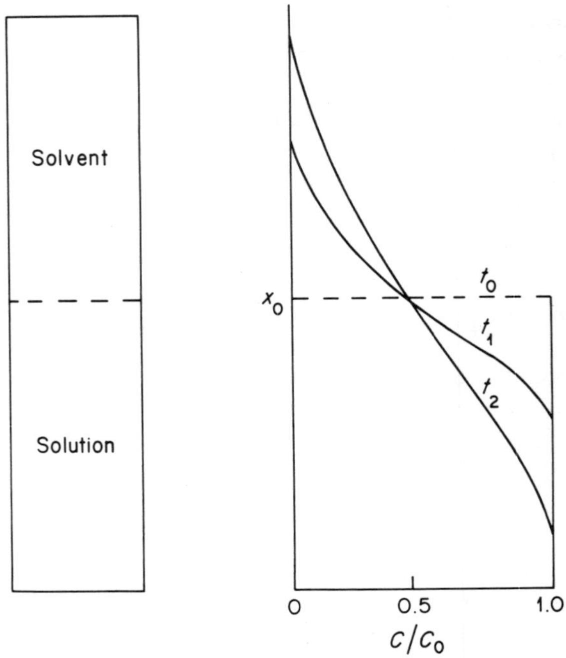

Figure 3.7 Diagrammatic representation of the experimental set-up for measuring diffusion by mass flow. Curves are drawn for three times, initial (t_0) and two succeeding records (t_1 and t_2)

Possibly the best-known solution is for a semi-infinite cell (Crank, 1956) in which the starting boundary is formed at $x = 0$, while the concentration at $t = 0$ is c_0, so giving

$$c_{x,t} = \frac{c_0}{2(\pi Dt)^{1/2}} \int_x^{\infty} \exp\left(-\frac{x^2}{4Dt}\right) \mathrm{d}x \qquad 3.16$$

The integral in eq. 3.16 is the complement of the error function and can be evaluated numerically. Measurements made from a diffusing boundary with schlieren optical systems (Lloyd, 1974, p. 21) which records $(\mathrm{d}c/\mathrm{d}x)_{x,t}$ versus x are best analysed through the derivative of eq. 3.16, i.e.

$$(\mathrm{d}c/\mathrm{d}x)_{x,t} = \frac{c_0}{2(\pi Dt)^{1/2}} \exp\left(-\frac{x^2}{4Dt}\right) \qquad 3.17$$

Equation 3.17 describes the familiar gaussian bell-shaped curve formed by a symmetrical boundary pivoting about a stationary point at $x = 0$ where $c/c_0 = 0.5$. The symmetry of the boundary makes it a suitable subject for recording by a Gouy interferometer (Gosting, 1956). From these basic formulae have been derived various subsets for analysing experimental boundaries. The most popular for analysing derivative profiles is the 'height/area' ratio calculated for several records collected over the experimental period and described by many authors discussing diffusion Detailed discussions of these analyses are not pursued further, and interested workers should consult Tanford (1961, p. 346) and van Holde (1971, p. 85) for details.

Several disadvantages were inherent in the classical approach to diffusion of macromolecules. The method is suited more to studying small molecules of the size of sucrose than to studying macromolecules. With sucrose, it is possible to record exact measurements within a relatively short period (measured in hours not seconds), while for macromolecules similar measurements may require days of recording. Some idea of the level of precision possible when dealing with molecules like sucrose can be gathered from the work of Gosting and Morris (1949). They were able to measure the concentration dependence of sucrose over a limited range of concentrations and show that this dependence decreased with increasing temperature (as expected from eq. 3.13). Similar studies on the diffusion of macromolecules in which temperature dependence can be noted are very rare, and this deficiency reflects the differences in the relative precisions. It is difficult to maintain accuracy while measuring the diffusion of gradients over long periods. Since the diffusion coefficient varies inversely as the third power of molecular weight, the time required to obtain acceptable changes increases from a few hours with sucrose to many days with virus. Isolating the gradient from physical vibrations or stabilizing the temperature to prevent density inversions for these long times becomes a major problem, especially when examining dilute solutions. Even when these practical limitations are overcome, there still remains the problem of estimating an average diffusion coefficient for a heterogeneous solute. Biologically pure samples of a macromolecule may still be heterogeneous physically if only because reversible associations can occur between the macromolecules (see Chapter 4). Although methods exist for processing results from diffusion which both detect as well as quantify heterogeneity, these generally depend on moment analyses similar to those described for the ultracentrifuge (eq. 2.38). The basic problem is to obtain an average diffusion coefficient from the diffusing boundary, and this alone limits the usefulness of boundary analysis for heterogeneous solutes. Hence, routine measurements of diffusion by this method are seldom useful in studying macromolecular interactions in solution. Only since quasi-elastic light scattering became routinely possible have these studies been attempted, so this technique is discussed more fully in Section 3.4.2.

One application of boundary analysis which still remains unsurpassed when studying biological macromolecules is the measurement of diffusion through its biological activity. Occasions when this method is ideal are in studies of impure samples of viruses and enzymes. Often the first indication of the presence of these particles comes indirectly from assays of activity. For these macromolecules the impure preparation and solvent are arranged as in Figure 3.7, with the preparation on top. For this method, a cell must be constructed which allows removal of samples from the solution at measured positions along the cell, and Trautman and Breese (1960) describe such a cell. After the solution and solvent have been put into contact and allowed to diffuse for a known long period, the samples are collected, assayed for activity, and c/c_0 calculated for several positions in the cell; c_0 is taken from a sample of the macromolecule stored for an equivalent period under identical conditions and assayed for activity. Each sample corresponds to an average spatial coordinate, so it is essential to make the divisions sufficiently small that the diffusion coefficient can be estimated using eq. 3.16. If this procedure is combined with a similar experiment using a preparative ultracentrifuge, where the sedimentation coefficient is again estimated from the movement of the activity, then a good estimate of the relative mass of the active molecule is possible. These two experiments can be performed on very impure preparations provided that hydrolysis of the macromolecule is inhibited during the analysis. The results give valuable information on the hydrodynamic properties of the active molecule rather than of the whole macromolecule. Using these techniques, Trautman and Breese (1960) were able to describe the approximate shape and relative mass of foot-and-mouth virus before the virus had been purified.

3.4.2 Quasi-elastic light scattering (QEL)

This physically based technique differs from conventional light scattering (Chapter 2) in that it assumes that the frquency of the scattered light is not the same as that of the incident beam (if it was, this would be elastic scattering); in fact, it depends on the differences for its operations. The easiest way to appreciate the operation is to call on everyday experiences with Doppler shifts of frequencies, in which acoustic waves are reflected from a moving object. These experiences show that the frequency of the reflected wave is not the same as that of the incident beam; in fact, the velocity (hence, frequency) of the reflected wave is the algebraic sum of the velocities of the incident wave and of the moving object. Extending this familiar example to the molecular level means extending the theory of scattering of electromagnetic fields from molecules as outlined in Chapter 2. The frequency of the scattered wave will be shifted from the incident frequency by an amount determined by the random velocity of the molecule at the time of scattering and by any uniform directional motion it possesses (for Brownian motion, this second contribution is zero when summed over sufficient molecules, but it is not zero in cases like electrophoresis). This shift can only be observed if the relaxation time of the molecular motion is longer than the relaxation time of the absorption of the incident photon and re-emission of the scattered photon, but since this process takes much less than a picosecond this is a reasonable assumption. Having used intuition to describe the scattering, the next question is how this small change in frequency can be measured, remembering that molecular motions have very long times compared with the period of light waves. For example, the frequency of visible light is about 10^{15} Hz, while the equivalent frequency for random movement of a macromolecule of relative mass in the range 10 000 to 50 000 dalton is between 10^5 and 10^6 Hz. These numbers show that any experiment which depends on measuring the incident and scattered frequencies separately and then subtracting the two will never give a useful difference. The only way to measure this small difference is directly through interferometry and recording the beats at the photo-surface of a sensitive detector (e.g. photomultiplier). Various experimental procedures have been described for carrying out this experiment but, before considering these in detail, it is necessary to convert the concept of frequency-shifted light scattering into mathematical equations which describe the underlying requirements. These discussions are general, but more specific points are covered in specialized monographs on light scattering (Chu, 1974; Berne and Pecora, 1976; Jamieson and Maret, 1973; Cummins and Pike, 1974).

Interference equations

If two sinusoidal fields (E_i) of angular frequencies ω_i ($= 2\pi f_i$) are mixed at time t, then the summed field (E_T) is given (Ford, 1972) as

$$E_T = E_1 \sin \omega_1 t + E_2 \sin \omega_2 t \qquad 3.18a$$

If the mixing occurs at a power-law detector (e.g. photomultiplier) which detects intensities and not fields, then eq. 3.18a must be converted to give intensities (I) using the Poynting theorem ($I \propto E^2$) to give

$$I_T = A(E_1 \sin \omega_1 t + E_2 \sin \omega_2 t)^2 \qquad 3.18b$$

where A is a constant which includes ampification of the photomultiplier as well as allowing for its relative efficiency (not every photon at the cathode gives an electron). Expanding eq. 3.18b using known trigonometrical relationships gives

$$I = A\{E_1^2 \sin^2 \omega_1 t + E_2^2 \sin^2 \omega_2 t + E_1 E_2[\cos(\omega_2 - \omega_1)t - \cos(\omega_2 + \omega_1)t]\} \qquad 3.19$$

Since photomultipliers do not respond to frequencies greater than 10^{10} Hz, the term containing $\cos(\omega_2 + \omega_1)$ becomes zero and the $\sin^2 \omega_i t$ are averaged to $\frac{1}{2}$, producing the following simple interference equation, i.e.

$$I = A[\tfrac{1}{2}E_1^2 + \tfrac{1}{2}E_2^2 + E_1 E_2 \cos(\omega_2 - \omega_1)t] \qquad 3.20$$

The value of the component ($\frac{1}{2}E_1^2 + \frac{1}{2}E_2^2$) is independent of frequency and becomes a direct-current (DC) baseline upon which is formed the beat frequency ($\omega_2 - \omega_1$). Thus, if two beams of light differing in frequency by ($\omega_2 - \omega_1$) are mixed on the cathode of a photomultiplier, an alternating field will be recorded at the anode having a frequency equal to ($\omega_2 - \omega_1$).

In a solution of macromolecules, the scattered light will contain many frequencies, some from the macromolecules and others from the solvent. Fortunately, the size of the macromolecules is so much larger than that of the solvent molecules that macromolecular motions are significantly slower than the solvent, so making its frequency considerably less, and the two can easily be separated. However, because the relative proportion of solvent molecules is so much greater than that of the macromolecules, scattering from these provides a substantial part of the DC signal collected from the scattered intensity (frequencies greater than 10^{10} Hz). This explains why simple measurements of beat frequencies from a fluctuating intensity require extensive processing before the macromolecular frequencies can be calculated.

Experimentally, the scattered light contains many frequencies, which means that the number of terms in a practical form of eq. 3.20 is increased to produce eventually the well known Fourier expression for a signal containing multiple frequencies. From this processed signal can be calculated the average diffusion coefficient. Visually the effect is to produce a speckled pattern from the scattered light, and this resembles theoretically the diffraction pattern produced in X-ray crystallography but differs in that Brownian motion is random so the relative positions of the speckles on the pattern are random (in crystals, the arrangement of scatterers is ordered so an ordered pattern emerges). Operationally (but not practically), it is necessary to adjust the aperture of the detector so that it just covers one of these speckles, and this is the coherence area discussed briefly on p. 131. The potential power of this method of measuring diffusion has been known for many years; in fact, Raman was one of the first to suggest that Brownian motion could be studied from scattering (Ramachandran, 1943). Subsequently, Raman (1959) observed motions of a film of milk using a filament lamp to illuminate the sample. However, it was not until the advent of stable lasers that the illumination was sufficiently steady and produced a sufficiently narrow bandwidth that the procedure became routinely possible. The first reported measurements were made with polystyrene spheres by Cummins *et al.* (1964) in which the laser beam was split. One beam was modulated at 18 Hz using a crystal and this was scattered from the solution. A second beam was modulated at 30 Hz and combined with the scattered light on the cathode of a photomultiplier. The power spectrum emerging from the anode was monitored in the vicinity of 12 MHz, and Cummins *et al.* (1964) were able to demonstrate a broadening of the spectrum caused by the random motion of the spheres. This and subsequent early experiments used wave analysers (a variable-frequency resonant circuit) to measure the broadening, but eventually this method was superseded by temporal autocorrelation which provided a more direct route to the diffusion coefficient.

Autocorrelation

Autocorrelation is a procedure used to compare a signal at time t with one at time $(t + \tau)$ but, since neither can the measurement of individual molecules be followed nor the time interval (τ) be made sufficiently small that the mathematical definition of autocorrelation is explicitly obeyed, each signal at time t and $(t + \tau)$ is an average—normally shown as $\langle E(t) \rangle$ or $\langle E(t + \tau) \rangle$. The relationship between autocorrelation and Fourier series is given through the Weiner–Khinchine theorem which states that the power spectrum and first-order temporal correlation function of any stationary ergodic signal form a Fourier transform pair (Pusey and Vaughan (1975) discuss this relationship in detail for diffusional processes, but the theorem itself is discussed in Davenport and Root (1958)).

Thus, for a power spectrum $(S(\omega))$, the relationship becomes

$$S(\omega) = \frac{1}{2\pi}\int_{-\infty}^{+\infty} |\langle E(t).E^*(t+\tau)\rangle| \exp[i(\omega - \omega_0)\tau]\,d\tau \qquad 3.21$$

where the exponential term is the trigonometrical series transformed using de Moivre's theorem for complex variables. It is seen from eq. 3.21 that if $\langle E(t).E^*(t+\tau)\rangle$ is independent of τ (constant intensity), $S(\omega)$ is a δ-function or spike at

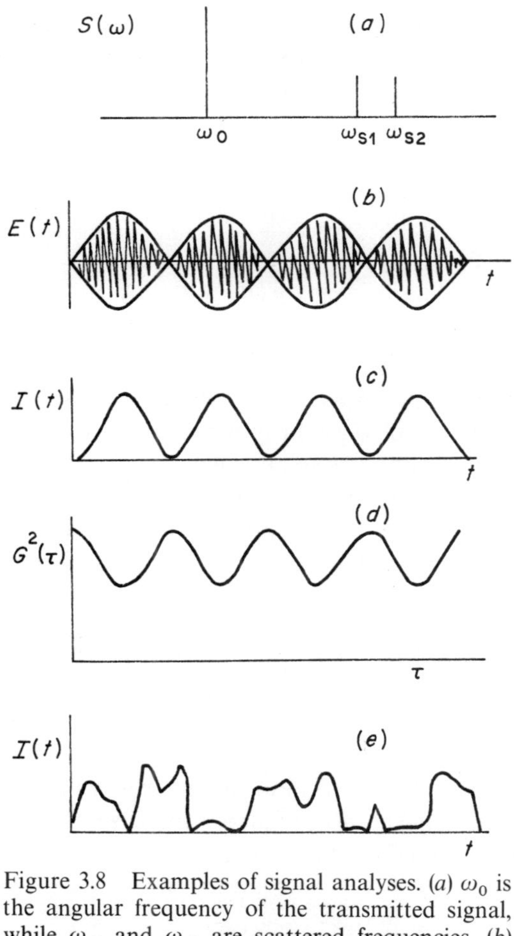

Figure 3.8 Examples of signal analyses. (*a*) ω_0 is the angular frequency of the transmitted signal, while ω_{s1} and ω_{s2} are scattered frequencies. (*b*) Scattered field showing the modulated envelope produced by beating ω_{s1} with ω_{s2}. (*c*) Corresponding modulated intensity. (*d*) Correlation function, $G^{(2)}(\tau)$ of the intensity. (*e*) Envelope of fluctuating intensity produced by beating of many fields. (Taken from Pusey and Vaughan, 1975, *Dielectric and Related Processes*, Figure 1. Reproduced by permission of The Chemical Society)

frequency ω_0, having zero width. However, if the correlation function decays with a characteristic coherence time T_c, then the frequency width of the spectrum (Δf) is given as $\Delta f = 1/T_c$. The term $\langle E(t).E^*(t+\tau)\rangle$ is the first-order correlation function normally written as $G^{(1)}(t, t+\tau)$ or simply $G^{(1)}(\tau)$, and often used in the normalized form

$$g^{(1)}(\tau) = \frac{\langle E(t).E^*(t+\tau)\rangle}{\langle I\rangle} \qquad 3.22$$

where $\langle I\rangle$ is the average intensity of the scattered field. Higher-order correlation functions can be defined, but in general only the second-order normalized function is important to diffusional studies. Here, the field strength is related by intensity to give the definitions

$$G^{(2)}(t, t+\tau) = \langle I(t).I^*(t+\tau)\rangle \qquad 3.23a$$

$$g^{(2)}(\tau) = \frac{\langle I(t).I^*(t+\tau)\rangle}{\langle I\rangle^2} \qquad 3.23b$$

$$= 1 + |g^{(1)}(t, t+\tau)|^2 \qquad 3.23c$$

These equations can be used to interpret results from light scattering but, before examining this, it is worth obtaining a general appreciation of scattering using the example from microwave radar, where the ideal scattered signal at the receiver will be Dopple-shifted by the moving aircraft to give a frequency ω_{s1} of constant intensity. For our purposes, it is better to consider the situation where two moving targets occur (resembling two independent molecules in a solution), when the second frequency is ω_{s2}. Clearly the detector gives a response which is modulated at the beat frequency

$$\Delta f = (\omega_{s1} - \omega_{s2})/2\pi$$

The result of these interferences is shown in Figure 3.8 taken from Pusey and Vaughan (1975). The output of the detector is given by $1/\Delta f$ (Figure 3.8) together with the corresponding $G^{(2)}(\tau)$ of the intensity calculated by autocorrelating the signal with itself at fixed time intervals (τ). Maxima are produced for $G^{(2)}(\tau)$ at an integral number of periods for harmonic signals, from which beat frequencies can be deduced. Increasing the number of targets increases the number of beats and, hence, the complexity of the signal, to give a record which would resemble the one given in Figure 3.8e.

Correlations and diffusion

Converting relaxation times measured by temporal autocorrelation to diffusion coefficients is not as difficult as it might seem. Ford (1972) gives an approximate derivation for an ideal solute undergoing Brownian motion. Light is not scattered from a medium consisting of perfectly uniform index of refraction. However, Brownian motion induces local concentration changes which, when averaged over the whole solution, gives $\langle c\rangle$, the average concentration. Locally, there will occur fluctuations in concentration (δc) which can be recorded through similar fluctuations in index of refraction (δn). If the index of refraction is only a function of local concentrations, then

$$\delta n(K, t) = \frac{\partial n(K, t)}{\partial c} \delta c \qquad 3.24a$$

$$\delta c(K, t + \tau) = \delta c(K, t) f(K, \tau) + \delta c' \qquad 3.24b$$

where K is the scalar wave vector of fluctuations in which the polarization of the scattered light is the same as the incident light ($= 4\pi n \sin(\theta/2)]/\lambda$ where n is the refractive index of the solvent). Now assuming that $\delta c(K, t)$ can be divided into the following two parts, (*a*) macroscopic fluctuations which obey the diffusion equation (δc) and (*b*) random processes having zero average but at any time equal to $\delta c'$ this equals the spontaneous generation of concentration fluctuations which in their absence gives $\delta c' = 0$), then $\delta c(K, t + \tau)$ is described by eq. 3.24b. Thus Fick's second law for diffusion in one dimension at infinite dilution (D°) becomes

$$\frac{\partial\, \delta c(x, t)}{\partial t} = D^\circ \frac{\partial^2\, \delta c(x, t)}{\partial x^2} \qquad 3.25$$

Inserting

$$\delta c(x, t) = \delta c(K, t) \sin Kx$$

for the concentration fluctuations followed by replacing $\delta c(x, t)$ in eq. 3.25 means that it is possible to solve for time dependence to give

$$\delta c(K, t) = \delta c(K, 0) \exp(-D^\circ K^2 \tau) \qquad 3.26$$

from which is deduced that

$$f(K, \tau) = \exp(-D^\circ K^2 \tau)$$

in eq. 3.24b while the first-order correlation function becomes

$$G^{(1)}(\tau) = \langle |\delta c(K, t)|^2 \rangle \exp(-D^\circ K^2 \tau) \qquad 3.27$$

The derivation outlined above is correct for ideal molecules having $\tau > 10^{-11}$ s. Equation 3.27 can be converted to give $g^{(1)}(\tau)$ (using eq. 3.21) and $g^{(2)}(\tau)$ (eq. 3.23) through the generalized definition in which $\Gamma = D_T K^2$, i.e.

$$g^{(1)}(\tau) = \exp(-\Gamma\tau) \qquad 3.28a$$

when

$$g^{(2)}(\tau) = 1 + f(A)|g^{(1)}(\tau)|^2 \qquad 3.28b$$

The term $f(A)$ compensates for destructive interference of wavefronts and is a function of the concentration of the molecules as well as their size; its value is determined by the coherence area (A_{coh}) employed in the experiment. This area is defined by

$$A_{coh} = \lambda^2 z^2 / s^2 \qquad 3.29$$

where z is the distance between the detector and the illuminated centre of the solution, and s is the illuminated width in the solution. If $A_{coh} \gg z\lambda/s$, then scattered fields are uncorrelated. Hence, it is necessary to adjust A_{coh} so that it equals the area of the detector (A_{det}), and if $A_{det} \ll A_{coh}$ then $g^{(2)}(\tau)$ decays from about 0.2 to 0.1 (Pusey and Vaughan, 1975) and $f(A)$ (eq. 3.28b) is approximately unity (theoretically, it ranges between 0.8 and 1.0). In practice, other instrumental factors influence $f(A)$, so it is not possible to define its value precisely. This means that it must be treated as a variable when fitting $g^{(1)}(\tau)$ or $g^{(2)}(\tau)$ against τ (autocorrelators measure $g^{(2)}(\tau)$ while wave analysers give $g^{(1)}(\tau)$) when

$$g^{(2)}(\tau)^{-1} = f(A)\exp(-2D_{\mathrm{T}}K^2\tau) \qquad 3.30$$

τ is a relaxation time and, since the argument of an exponential must be dimensionless,

$$\tau = (D^\circ K^2)^{-1} \approx \lambda^2/D^\circ$$

(since $K = [4\pi n \sin(\theta/2)]/\lambda$ where n is the refractive index of the solvent). From this, it can be surmised that τ is the time taken for the molecule to move a distance equal to the wavelength of the incident light (λ). This simple interpretation indicates that, in order to follow high-frequency motions (with small amplitudes), it is necessary to use short-wavelength electromagnetic radiation (for a more detailed discussion of the physics of light scattering, see Pusey and Vaughan, 1975).

Occasionally, one meets confusion in the literature as to whether diffusion coefficients measured by QEL equate physically with those measured through either mass transfer (Section 3.4.1) or redistribution of isotopically labelled molecules (in the latter method, a boundary is formed between the solutions of labelled and unlabelled molecules at the same weight concentrations and the temporal distribution of the labelling followed). The argument following from eq. 3.25 shows that QEL measurements do relate to concentration gradients, so it seems that identical diffusion coefficients would result from QEL and mass transfer measurements. However, in the labelling experiments, no mass transfer of solute or solvent is involved, and it is realistically assumed that the chemical potentials of the molecules are not changed by labelling. (This could be challenged on theoretical and experimental grounds for small molecules, since under some circumstances the stable isotopes of inert gases such as helium and argon can be separated through liquefaction, while isotopes of other gases can be separated by diffusion; in these cases, the proportional changes in masses of the isotopes are large compared with labelling macromolecules.) Intuition suggests that, in the isotope experiments, the positions of macromolecules could be interchanged with no apparent nett exchange of solvent and one is following the spatial rearrangement of labelled molecules among identical molecules. Basically, the method resembles the procedure where the observer measures through a microscope the random walk (x, eq. 3.7) of, say, a bacterium or large particle and then calculates a diffusion coefficient. Thus, since no gradient of weight concentration is involved, the coefficients from isotope experiments do not equate physically with those measured through mass transfer. Unfortunately, there are exceptions to this rule when applied to QEL. Under some circumstances it is possible to conduct experiments where the results stem essentially from following the movements of single molecules. This condition occurs when the wavelength of the incident radiation is similar to or smaller than the average dimensions of the macromolecule. Experimentally, this arises when studying either large particles with visible light or moderately sized macromolecules with short wavelengths (e.g. X-ray scattering from solution; for an introduction to this method, see Eisenberg, 1976); the problem lies in adjusting the coherence area (eq. 3.29) to match the experimental coordinates. It is also possible to satisfy this condition for moderately sized macromolecules when using visible light if back-scattering is measured rather than forward-scattering (the normal mode). In back-scattering, the scattering vector is formed from a sum of two vectors, so making the operational vector (which contains the wavelength of the incident radiation) appear to be very large; in forward-scattering, this vector is formed from the difference between two vectors. These arguments show that some care is required when comparing diffusion coefficients measured by different methods, and it is important to realize that physically the interpretation of the accompanying phenomena (such as concentration dependence; Batchelor, 1976) do not yield the same theoretical values for experiments involving mass transfer as with identifiable molecules.

Diffusion and polydispersity

In deriving the relationship between temporal autocorrelation and diffusion, it has been assumed that only one relaxation time is present, corresponding with translational diffusion of a single species. Generalizing these relationships so that they include rotating or flexing as well as polydispersity for translational diffusion produces the impractical situation where eq. 3.30 contains an infinite sum of exponentials (as in eq. 2.26). It is possible in some cases (where the range of molecular masses differs by several hundred thousands) to separate translational diffusion from other diffusion processes by restricting the frequency range studied. Employing this process generally can be dangerous and may produce distorted results, but by studying the angular dependence of scattering it is always possible to separate rotational from translational diffusion. In all cases, the autocorrelation function varies with the angle between the incident and scattered beams and, since the records at 90° are formed from the sum of translational plus rotational, while only translational diffusion is measured at 0°, it is possible to separate the two forms of diffusion. The method used to separate these is the same as that used for intensity light scattering, namely through a Zimm plot (Chapter 2, p. 89). If the scatterers of a polydisperse mixture are small compared with K^{-1} (i.e. averaged dimensions less than $0.05\,\lambda$) and all have the same polarizability per unit volume, then the average translational diffusion coefficient at infinite dilution is a Z-average ($\bar{D}_Z$) defined as

$$\bar{D}_Z = \frac{\bar{\Gamma}}{K^2} = \frac{\Sigma_i N_i M_i^2 D^\circ_{\mathrm{T},i}}{\Sigma_i N_i M_i^2} \qquad 3.31\mathrm{a}$$

$$= \frac{\Sigma_i c_i M_i D^\circ_{\mathrm{T},i}}{\Sigma_i c_i M_i} \qquad 3.31\mathrm{b}$$

where N_i is the number of molecules per litre (concentration c_i in $\mathrm{g\,l^{-1}}$) having molecular weight M_i and translational diffusion coefficient $D^\circ_{\mathrm{T},i}$. Combining $\bar{D}^\circ_Z$ with a weight-average sedimentation coefficient (Section 2.1) produces a weight-average molecular mass through the Svedberg equation (eq. 2.16) (Kinell, 1959). QEL is the only experimental method which yield $\bar{D}^\circ_Z$ directly.

Association between molecules can be detected by the characteristic concentration dependence of D_T; this follows the form shown in Figure 4.4 (myosin) and Figure 3.5 (glutamate dehydrogenase). This type of dependence must be separated from that caused by non-ideality, which has been discussed earlier in this chapter. When chemical interaction between macromolecules occurs, the value of the average diffusion coefficient calculated at each concentration will be $\bar{D}_Z$ only if the data are processed correctly using one of the procedures discussed by Pusey and Vaughan (1975).

Large particles

A more complex problem arises with non-interacting particles having average dimensions very much greater than λ (e.g. TMV), especially if they are asymmetric and flexing. When these particles are aligned parallel to the incident wavefront, local interference between scattered wavefronts decreases the intensity of scattering (Chapter 2, p. 88); when aligned perpendicularly, no interference occurs but some phaseshift would be produced. In this case, eq. 3.31a must be modified to include rotational (D_r) time constants, giving

$$g^{(1)}(K,\tau) = \frac{\Sigma_i N_i M_i^2 P_i(K)\exp(-D_i K^2 \tau)}{\Sigma_i N_i M_i^2 P_i(K)} \qquad 3.32$$

where $P_i(K)$ is the particular scattering factor at angle θ for the ith diffusion type (i.e. translational and rotational). Pecora (1968) has considered this problem and shown that $g^{(1)}(\tau)$ may be written as

$$|g^{(1)}(\tau)| = A_0(K)\exp(-D^\circ_\mathrm{T} K^2 \tau) + A_1(K)\exp[-(D^\circ_\mathrm{T} K^2 + 6D_\mathrm{T})\tau] + \text{other terms} \qquad 3.33$$

where A_0 and A_1 are K-dependent amplitudes ($A_0 \gg A_1$), and so this equation could be fitted to data from large particles. It can be inferred from eq. 3.33 that translational diffusion coefficients for large asymmetric particles can only be obtained from measurements of forward-scattering. Thus, if measurements from QEL for bacteriophages and similar particles are to be employed in estimating relative molecular masses through the Svedberg equation (eq. 2.16), then corrections for the angular dependence of the scattering must be made (for example, Zimm plot, Figure 2.11). Pusey (1973) has developed a further procedure for analysing mixtures of large or small particles based on the idea that $g^{(1)}(K, \tau)$ (eq. 3.32) can be expanded about $\exp(-\bar{D}^{\circ} K^2 \tau)$ where

$$\bar{D}^{\circ} = \int DP(K)\mathrm{d}D$$

and equals the Z-average diffusion coefficient measured at small angles. The result of this expansion gives an average diffusion coefficient ($\bar{D}$) and dispersion ($D^2 - \bar{D}^2$), to give

$$\ln[g^{(1)}(K, \tau)] = -(D^{\circ})^2 K^2 \tau + [(D^2 - \bar{D}^2)K^4 \tau^2 + \ldots] \qquad 3.34$$

Again this equation can be fitted to data from polydisperse solutions and the dispersion can be compared directly with the dispersion calculated for the sedimentation coefficients (Baldwin, 1957a, b, and discussed in Section 2.4.1).

A more detailed analysis of light scattering from filamentous viruses (experimental and theoretical) has been given by Loh and colleagues (Loh *et al.*, 1979; Loh, 1979). These experiments show that, in addition to translational and rotational diffusion of whole particles, a further fast motion can be observed which is explained through the presence of bending modes of the whole rod. This bending may be coupled to the rotational diffusion and produces an enhancement of the fast amplitude of the scattering. Bending of anisotropic particles is to be expected (see Section 3.2.3), but this appears to be the first experimental observation of its recording through measuring the amplitudes of scattered light.

Experimental

The theoretical considerations illustrate the possible applications of QEL for measuring molecular motions. Realizable measurements can only be made using specialized equipment such as wave analysers or autocorrelators, and, since the latter equipment has become standard for this technique, the description of experimental techniques will be restricted to this procedure, namely temporal autocorrelation (method 3). Confusion between procedures can arise because a wide-ranging terminology is used to describe the experiments. Practically, there are three configurtions possible for studying QEL:

(1) The incident beam is split and each beam modulated before mixing at the cathode (see p. 128). This system is universally called heterodyning.

(2) A stationary scatterer (e.g. plastic bead held on a rod) is placed in the solution at the centre of the incident beam. This scatterer provides an intense local oscillator with which the scattered light beats at the photocathode. This is called homodyning.

(3) The scattered light is allowed to beat with itself (as opposed to beating with other oscillators as in (1) and (2)) and the noise signal from the photomultiplier is analysed to extract the periodic components. This is called either self-beating

spectroscopy or the direct intensity fluctuation method and is the method most used in experiments on diffusion. Further terms for this process are intensity fluctuation spectroscopy (IFS) and photon correlation spectroscopy (PCS). Pusey and Vaughan (1975) prefer to retain IFS as a general term not carrying connotations about methods used for signal processing, and they retain PCS for digital correlation processes.

All three procedures follow differing routes to measure broadening of the Rayleigh line. (The Rayleigh line is the spectral line of the scattered light and would be a δ-function centring on the frequency of the incident light for elastic scattering; the spectral broadening is produced by the random motion of the molecules.) It is worth restating here that through QEL a new chapter in biochemistry is opened, since it is now possible to observe almost individual motions at the molecular level. The theory is sound and largely proven at the atomic level, but macromolecular applications providing new and hitherto unmeasured parameters are limited: results are largely confined to proving consistency with previous measurements (see Section 3.32).

Temporal autocorrelation

A detailed description of the composition of an intensity fluctuation correlator is given by Pusey and Vaughan (1975) while Pusey *et al.* (1974) provide a simple diagrammatic representation of the stages occurring in processing the ideal signal (reproduced in Figure 3.9). During each sampling interval τ, the number of

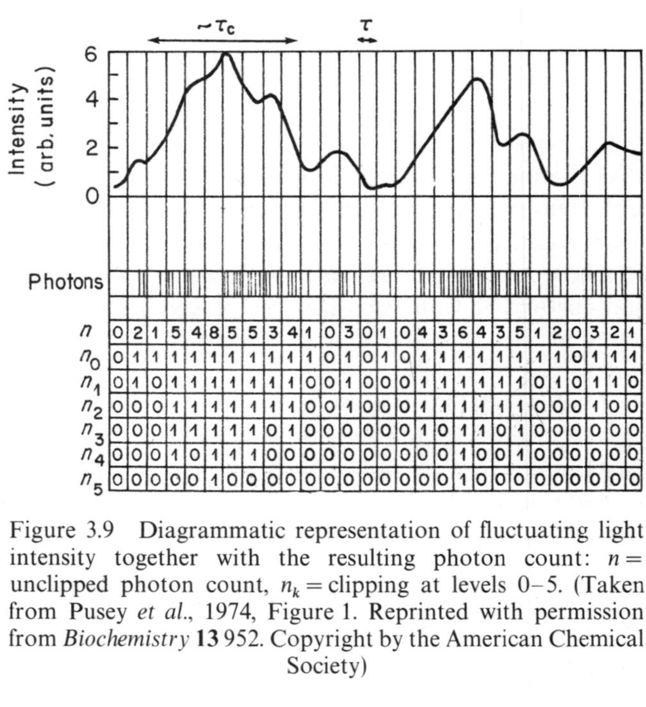

Figure 3.9 Diagrammatic representation of fluctuating light intensity together with the resulting photon count: $n=$ unclipped photon count, $n_k=$ clipping at levels 0–5. (Taken from Pusey *et al.*, 1974, Figure 1. Reprinted with permission from *Biochemistry* **13** 952. Copyright by the American Chemical Society)

photons recorded by the photomultiplier are accumulated in a counter (n, Figure 3.9). Thus, for the example given in Figure 3.9, the equipment would require as many counters as there are time intervals (m) and the 'full' autocorrelation would involve calculating the partial products

$$\sum_{i=m}^{N} n(t_i)n(t_{i-m}) \qquad 3.35$$

for the range of delay times $m\tau$ where $m = 1, 2, 3, \ldots$ (N is the number of counters or channels built into the autocorrelator), and adding these to the previous sums. Calculating cross-products of numbers is a slow process when precision is required, and hence a modified operation is employed in which the signals from the photomultiplier are 'clipped'. This involves setting an electronic trigger to a convenient level (k) so that

$$n_k(t_i) = 1 \qquad \text{if } n(t_i) > k \qquad 3.36a$$

$$n_k(t_i) = 0 \qquad \text{if } n(t_i) \leq k \qquad 3.36b$$

These clipped levels are given as n_0, n_1, n_2, etc., in Figure 3.9. Thus, provided that the summed counts in each channel are large (as well as making $N\tau/\tau_c$ large), any one sum will be close to the average value obtained using eq. 3.35 and will be equal to $g^{(1)}(\tau)$ (eq. 3.28). Clipping the input has the great advantage that multiplication in base 2 is equivalent to adding '1' or '0' to the sum depending on whether the multiplicand is '1' or '0'. Thus, the correlator is fast (the limit on τ is about 30 ns) and accurate, while the electronic complexity is reduced, since the instrument can be built from digital electronic units. Practically, the 'clipped' correlator consists of a 'clipper' feeding counters and shift-registers. The outputs of these registers are connected through 'AND' gates to the N counters. The second input to the 'AND' gates are also connected to the shaped and clipped output from the photomultiplier. The time interval is controlled by an accurate clock and, at each interval, pulses from this clock not only shift the contents of the shift-register towards higher times but also activate the counters so that they respond to the outputs from the 'AND' gates. Thus, at each clock-tick, if any individual cell of the shift-register contains a '1' and a '1' comes from the photomultiplier, then '1' is added to the counter connected to the 'AND' gate' If either the cell or the photomultiplier gives '0', then nothing is added to the counter. Additional counters store the total time and number of photons recorded; the latter is required to normalize the contents of the counters (equivalent to $\langle I \rangle^2$ in eq. 3.23b). The rest of the equipment consists of a stable laser (argon-ion at about 50 mW is suitable) focused to illuminate a cuvette held in a water bath maintained at the required temperature and situated at the centre of a goniometer table (two-dimensional). The photomultiplier plus slits, etc., are mounted on a rigid but movable arm that pivots about the centre of the table. The arm can be placed at angles relative to the incident beam and it is essential that the cuvette be placed accurately at the pivot point of this arm. Fuller details of the equipment and methods of handling samples so that dust is reduced are given in many papers (e.g. Pusey *et al.*, 1974; Pusey and Vaughan, 1975; Cummins and Pike, 1972), so it is not repeated here. In common with most techniques, there are optimum operating conditions for making successful measurements of QEL. Possibly the most important are (*a*) controlling the temperature accurately to about $\pm 0.1°C$ (a periodic fluctuation in temperature can produce spurious results which resemble very slow diffusion processes which are really caused by density inversions in the cuvette and these are measured as low-frequency signals), (*b*) ensuring that the correct coherence area is produced by adjusting the aperture in front of the photomultiplier, (*c*) ensuring that the correct illumination is obtained to give optimal count rates for the required signal—to noise ratio, (*d*) adjusting the sampling time so that it matches the number of channels in order to accommodate the complete curve of the relaxation, and (*e*) finally adjusting the 'clip level' to match the photon flux. Guidance for setting these instrumental parameters are given by Chu (1974) and Pusey and Vaughan (1975).

With PCS it is possible in principle to measure both translational and rotational diffusion coefficients by recording the autocorrelation at a series of angles. Practically, the problem reduces to one of finding a suitable statistical procedure which separates the individual components from the complex record. The difficulties encountered in this fitting are determined by the accuracy of the method, because the useful parameters all appear as arguments to exponentials (as in eq. 3.33), making accurate estimates of many parameters difficult, or even impossible in many cases. Nevertheless, for monodisperse samples or samples containing a known number of (but few) components, valuable information about molecular motions can be gathered (see, for example, the assembly of T2 phage, p. 122, and polymerization of glutamate dehydrogenase, p. 121).

Other correlation procedures

The use of temporal autocorrelation in studying random motions is not restricted to light scattering: movement of fluorescent particles can be followed by this procedure (Magde, 1976). Temporally averaged motions of these particles provide estimates of parameters similar to those determined from light scattering. Also, the correlation procedure has been extended to methods which in principle can measure diffusion coefficients and electrophoretic mobility in the same relatively short experiment (true Doppler scattering). In this procedure, the measurement of scattering is coupled with the analysis of biased intensity fluctuations (Lim *et al.*, 1977). Combining the measurement of random movements with mass migration along one coordinate changes the relationship between $g^{(2)}(\tau)$ and the exponential relationship for diffusion (eq. 3.30) into one containing the diffusion term plus an additional term which describes the directed movement of the mass. The effect of this extra term is to give a peak in the relationship between $\ln[g^{(2)}(\tau) - 1]$ and τ (eq. 3.30) whose displacement in the temporal axis is a measure of the velocity of the mass movement. Measuring electrophoretic mobility by this procedure is not chemically ideal because, in order to produce sufficient mass transfer in the timescale of the autocorrelator, it is necessary to pulse the solution with relatively high voltages for periods up to one second. The high power makes it difficult to employ reversible electrodes (e.g. Ag/AgCl) and, at low ionic strengths, electrolysis of the water occurs. At its worst, this produces bubbles of O_2 and these are a disaster for light-scattering experiments. Even without the bubbles, the electrolysis must generate a pH gradient between the cathode and anode after each pulse, another undesirable condition for electrophoretic experiments. One ingenious application of QEL has been in following chemotactic responses of micro-organisms or other rhythmic movements such as cilia movements; again, this uses similar theory to that employed for electrophoresis, producing non-exponential decays of $g^{(1)}(\tau)$ (Schaefer and Berne, 1975).

Raman scattering

QEL is usually considered separately from Raman scattering, although the division is largely a matter of convenience rather than theory. Practically, the difference lies in the fact that with QEL the Rayleigh line is broadened and does not break up into defined spectral features, whereas in Raman scattering it does. Both phenomena can be treated theoretically through classical and quantum mechanics, although it is usual to discuss QEL through classical theory and Raman by considering the quantum energies of the scatterers. A detailed analysis of Raman scattering is given by Chu (1974) and only the outline of the principle is given here in order that the information gleaned by the process for macromolecules can be discussed briefly.

Previous discussions suggest that if any substance present in the gaseous, liquid or solid state is exposed to monochromatic radiation, then radiation scattered at 90° to the incident beam contains a range of frequencies whose values are determined by the activity of the substance. Unlike QEL, Raman scattering is concerned with measuring the frequencies of oscillatory groupings within the molecule and the shifts in frequencies are correspondingly larger than for diffusion and involve higher-energy transitions. The accepted mechanism for producing Raman scattering is that, when the molecule absorbs the incident radiation (this need not be a resonance absorption, although resonance Raman is more sensitive than non-resonance scattering), it is raised to a higher quantum level; it then emits radiation as it falls back to a level usually situated between the initial and higher levels. The energy of the Raman radiation ($hc\Delta\nu$, where h is Planck's constant, c is the speed of light, and $\Delta\nu$ is the difference in frequency between incident and Raman radiation) is then the difference between the initial and final state. In some cases, the final level has less energy than the initial state, so the frequency of Raman radiation is greater than the incident radiation, producing the so-called anti-Stokes line. This qualitative description of the process hides a range of detail but shows that, in general, Raman scattering provides information about molecules which is many respects is similar to that obtained from vibrational and rotational spectroscopy (ultraviolet (UV), visible, and infrared (IR) spectroscopy). Results can be interpreted in at least two levels of detail: (*a*) measurement of energies of interaction between atoms in neighbouring group involving the assignment of low-energy transitions to all frequencies, followed by calculating the appropriate wavefunctions, or (*b*) an analytical mode where difference in spectra for a chemical group are measured over a variety of molecules of differing but known conformations in order to correlate frequency shiffs with the environment of the groupings. These correlations can subsequently be empirically equated with shifts caused by conformational perturbations in complex molecules carrying unknown conformations. Generally, the second method is the one applied to macromolecules in solution: difficulties in calculating wavefunctions for solvated groups make the first approach less exact.

Possibly the first detailed analysis of the Raman spectrum of a protein (using the second method) was made by Lord and Yu (1970a) for native lysozyme. They were able to describe the number and local conformation of –C–S–S–C– groupings and C–S– (as in methionine) arising from – S–S– and –C–S– stretching frequencies which appear in the region 500 cm^{-1} and 725 cm^{-1} (a wavenumber in cm^{-1}, representing the number of wavelengths per cm, is the reciprocal of the wavelength given in cm). Similarly, from measurements of the lines of the amide group (1660 cm^{-1} and 1260 cm^{-1}), it was possible to obtain information about the peptide –CONH groups. The results suggest that the components of the structural backbone of lysozyme exist in three distinct structural conformations. A separate study also made by Lord and Yu (1970b) on native ribonuclease and α-chymotrypsin has a better resolution in the spectrum than for lysozyme, and showed that the conformation of –C–S–S–C– cross-links differ between the proteins. Extending this technique to studies on the effects that denaturation has on conformations is an obvious step, but results using Raman scattering are so far limited. As can be seen from this brief review, the wealth of detail obtained from Raman scattering equals that from other spectroscopic methods, making it a method for the future and requiring specialized knowledge to interpret. It remains to be seen how this technique will develop and supplement existing methods for studying macromolecules in solution. One major point in its favour is that it is one of the few methods which give details of conformation of groups which are comparable with results from X-ray diffraction studies. In this respect, it may span the interval between crystals and solutions when determining detailed structures of solvated macromolecules.

3.4.3 Dielectric relaxation and other methods for measuring rotary diffusion coefficients

The methods described in Sections 3.4.1 and 3.4.2 measure the random motions of macromolecules in solution through what are essentially inert procedures. These methods are not particularly sensitive for measuring rotational diffusion coefficients. Greater sensitivity can be achieved if the macromolecule is perturbed from its equilibrium arrangement by placing it in either a pulsed or sinusoidal electromagnetic field. Then, through a suitable monitoring system, it is possible to record the motions of the macromolecules after or during the application of the field. The basic theory of the orientation of molecules in an electric field is considered by Tanford (1961, p. 105) and van Holde (1971, p. 202): more comprehensive reviews are given by Pethig (1979) and Grant *et al.* (1978). Thus, only sufficient theory is given here to make it possible to follow the results.

Proteins and nucleic acids carry charges even at their isoelectric points, while the residues have asymmetric polarities which endow the polymer with a large permanent dipole moment whose value is determined in part by the pH of the solvent. Thus, placing a solution of these macromolecules in a polarized electric field causes the macromolecules to align themselves along the field. This ordered arrangement is opposed by the Brownian rotary motion (as discussed earlier). If the frequency of the alternating field is sufficiently high, then insufficient time remains between cycles for the molecules to rotate in sympathy with the field, and this change from ordered motion to random arrangement while under the influence of the field produces a marked break in the optical properties, and this can be used to measure the rotary diffusion coefficient.

A molecule subjected to an external field not only has its own intrinsic polarizability but also carries an induced polarizability (α_i) (see Chapter 1 for discussions of polarizability of molecules). Thus, in a field, the equation relating dielectric constant (ε) with polarizability (α) becomes

$$(\varepsilon - 1)/(\varepsilon + 2) = \tfrac{4}{3}\pi N(\alpha_i + \mu^2/3kT) \qquad 3.37$$

where N is the number of molecules per cm^3 and μ is the orientation dipole moment (defined in eq. 2.2). The second term in eq. 3.37 is a special case of the generalized polarizability equation given in eq. 1.36 and 1.37. Equation 3.37 could be used to determine the permanent and induced dipole moments of macromolecules, which in turn describe electric dichroism and electric birefringence (Kerr effect): both of these phenomena are induced through asymmetry of the electric charges on the molecule. Dichroism is measured by comparing the proportion of polarized light absorbed by a molecule orientated at angle θ with respect to the beam, compared with that absorbed when the molecule is aligned parallel to the beam ($A_\theta/A_{\parallel}$). Since refraction (n is refractive index), absorption, and dielectric properties are all functionally related ($n^2 \propto \varepsilon$), orientated asymmetric macromolecules will display different refractive indices for light polarized parallel to and perpendicular to the electric field. The relationships between these properties are complicated and usually derived through a quantum-mechanical approach, but for a rodlike molecule of mass m the refractive index differences at concentration c compared with the solvent (Δn) is given by

$$\frac{\Delta n}{c} = \frac{2\pi\varepsilon^2}{15Nm}\left(\frac{\mu}{kT}\right)^2(\alpha_{\parallel} - \alpha_{\perp}) \qquad 3.38$$

where $\alpha_{\parallel}$ and $\alpha_{\perp}$ are the polarizabilities parallel to and perpendicular to the electric field, E. Equation 3.38 is idealized and mainly serves to demonstrate the damping introduced by increasing the mass of the molecule. A classical derivation of this equation would be from equations for a damped harmonic oscillator in which the rotary acceleration of the molecule is inversely proportional to the mass (eq. 3.6). This aspect is not pursued further here because applying these idealized models to studies of proteins and nucleic acids in aqueous solution is difficult but is described by Pethig (1979). The presence of excess polar water molecules around the macromolecule modifies the local field to such an extent that the Debye equation (equivalent to eq. 3.41) may no longer hold. Furthermore, the fact that most solvents for biological macromolecules are efficient conductors of electricity means that the electric field strengths that can be applied are limited by the level of ohmic heating that the molecules can withstand (see Grant *et al.*, 1978, for details of the experimental procedures).

Much of this basic theory was developed in the later 1930s and early 1940s and the work carried out during this period by Oncley and co-workers showed that useful information on the shape of macromolecules in solution could be gleaned by measuring dielectric properties of solutions (for a review of this work, see Oncley (1942) and Edsall and Wyman (1958)). They expanded eq. 3.37 so that it included an empirical constant for local modifications of the electric field caused by both the solvation with water and the local high concentrations of counter-ions (Dintzis *et al.*, 1954). Applying these theories to interpret results for ionic macromolecules obtained through the Kerr effect showed that for certain macromolecules the orientation in an electric field was determined mainly by the polarization associated with ionic motions (O'Konski, 1960). Modern thoughts on this polarization are discussed by Pethig (1979).

Rotation of serum albumin

Moser *et al.* (1966) were able to glean information on the shape and interactions of bovine serum albumin in dilute salt from both the decay of birefringence occurring after pulsing the solution with 5000 V for 1.6 μs and from the change in dielectric constant occurring with increasing frequency of an applied field (up to 0.5 MHz). The decay occurs because first the pulse aligns the asymmetric albumin in the field, so changing the absorption of polarized light relative to that occurring with randomized orientations (in the manner described by eq. 3.38). When the field was removed, Brownian movement of the molecules returned them back to their original disordered state. The process was monitored by passing the polarized light through the solution and recording the absorption after pulsing (in their experiment, it took less than 1 μs to relax back to the random state). The relaxation time for this decay (τ_n, that is, the time taken for the absorption to return to 1/e of that it had before the pulse) is related to the rotary diffusion coefficient, D_r, by (Tinoco, 1955)

$$D_r = 1/6\tau_n \qquad 3.39$$

From the dielectric measurements at various concentrations and temperatures, a curve for the results extrapolated to infinite dilution was constructed

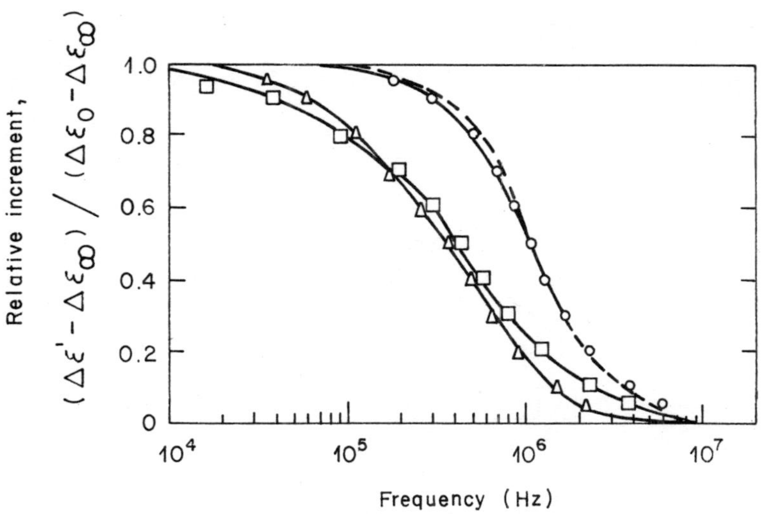

Figure 3.10 Relative dielectric increments (eq. 3.41) versus frequency for deionized bovine serum albumin at infinite dilution: o, monomer at 25°C; Δ, monomer at 1°C; □, dimer at 25°C. (Taken from Moser *et al.*, 1966, Figure 5. Reprinted with permission from *J. Phys. Chem.* **70** 741. Copyright by the American Chemical Society)

(Figure 3.10) and the relaxation time for the asymmetric molecule was calculated from the two-term Debye equation (Debye, 1929), i.e.

$$\frac{(\Delta\varepsilon' - \Delta\varepsilon_\infty)}{(\Delta\varepsilon_0 - \Delta\varepsilon_\infty)} = \sum_{i=1}^{2} \frac{A_i}{1 + (\omega\tau_i^2)} \qquad 3.40$$

A_i is the amplitude for the ith species and this is related to the Kerr constant. $\Delta\varepsilon'$ is the dielectric loss at infinite dilution calculated from a single-term Debye equation, i.e.

$$\frac{\Delta\varepsilon'}{c} = \frac{\varepsilon_\infty}{c} + \frac{(\Delta\varepsilon/c) - (\Delta\varepsilon_\infty/c)}{1 + (\omega^2\tau_\varepsilon)^2} \qquad 3.41$$

In both equations, $\Delta\varepsilon_0$ and $\Delta\varepsilon_\infty$ are the specific dielectric increases at frequencies below and above the dispersion region (Figure 3.10), while

$$\Delta\varepsilon'/c = (c_p - c_s)/cc_0$$

in which c_p, c_s and c_0 are the complex electrical capacities of a cell filled with protein, solvent, and pure water (cell constant), respectively. τ_i are the relaxation times, where $\tau_{\varepsilon 1} = 3\tau_n$ (eq. 3.39). Two important conclusions were drawn from these experiments:

(1) Bovine serum albumin is asymmetric and has two rotary diffusion coefficients, one considerably smaller than the other (the major one is about 2.2 μs at infinite dilution).
(2) There is a marked concentration dependence in the values of this major

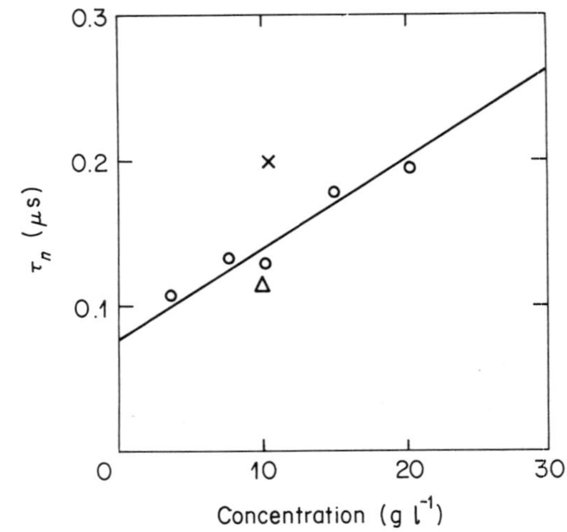

Figure 3.11 Concentration dependence of relaxation times determined by relaxation of birefringence for bovine serum albumin (monomer) at 25°C (o), together with values from other investigations ($\times$ and Δ). The straight line is obtained by linear regression. (Taken from Moser *et al.*, 1966, Figure 6. Reprinted with permission from *J. Phys. Chem.* **70** 741. Copyright by the American Chemical Society)

rotary diffusion coefficient. Figure 3.11 gives the extent of this dependence, and first appearances suggest that both the sign of the slope (negative for D_r, see eq. 3.39) and its magnitude are similar to those for translational diffusion of bovine serum albumin (Figure 3.1). However, in the case of translational diffusion measured at low ionic strengths near the isoelectric point for albumin, the concentration dependence showed a positive slope. Thus, although the macromolecular interactions described by eqs. 3.12 and 3.13 produce similar magnitudes for the two types of motion, the difference in signs of the slopes at low ionic strengths shows that the sum of the interactions are attractive for rotation and repulsive for translation. This marked difference obviously requires confirmation but, if it is supported, it suggests that, for high concentrations of protein at low ionic strength, rotational motions pull the molecules closer together while translational motions repel one another: at normal ionic strengths, the forces of translational motion are attractive.

Laser excitation

These experiments illustrate the problems associated with measuring rotations using pulsed electric fields. It is difficult to include salts in the solvent, and even when these are included they must be in low concentrations (in order to reduce heating), so the interpretations become uncertain. For example, the fluctuating movement of the counter-ions contributes to the total dispersion, so these effects must be subtracted from the measurements before the true macromolecular motion can be measured. For these reasons, measurement of circular dichroism became more popular for studying rotational diffusion coefficients. Here the absolute conversions only become possible if the magnitude of the electric fields is known, and this cannot be measured at molecular levels: the presence of water and counter-ions modify the microscopic levels. Hence, the technique has only been applied empirically to macromolecules.

With the advent of powerful lasers, it has been possible to re-examine the use of dichroism for studying the motions of macromolecules by producing powerful electric fields from the electric vector of the laser beam, with the minimum of heating. Interpreting this electrically induced dichroism is complicated because the high field strengths lead to non-linear optical effects through multiphoton absorbances, but it is an interesting development which it is too early to assess. Jennings and co-workers have developed this technique (Plummer and Jennings, 1969; Jennings, 1973, 1974, 1975), and their conclusions suggest that useful information may be forthcoming from the wider application of this method.

Fluorescent lifetimes

A further optical method enlists the knowledge that, when an atom or molecule in its quantum ground state absorbs radiation of short wavelength, it is raised to an excited state as a result of electronic transitions. If, while in this excited state, the molecule does not either decompose or collide with another atom or molecule, then within a very short time (between 10^{-12} and 10^{-8} s) it returns to a lower level and loses the energy by emitting radiation. This is called fluorescence, and the frequency of the emitted radiation corresponds to the difference in energy between the initial and final levels. If the exciting radiation is polarized, then absorption of energy will be most likely for molecules with transition dipole moments parallel to the direction of polarization. If the relative position of this orientation does not change during the period between absorption and emission of the photon, then the fluorescence remains polarized; if the molecule rotates during this period, the fluorescence is depolarized. Many biological macromolecules carry chromophores absorbing in the visible or ultraviolet regions of the spectrum, and thus it is possible to use this method to explore the rotation of molecules bearing the chromophores. The apparatus used for this work consists of a source of monochromatic radiation (usually ultraviolet) filtered through a polarizer and passing through a cuvette containing the sample. Situated at 90° to the incident beam is an analyser and a second monochromator, for measuring the angle of rotation of polarization of the emission; this is followed by the detector. The sample is pulsed with light of duration no longer than 10^{-9} s. This is then followed by recording the decay of the intensity at positions parallel to ($I_{\parallel}$) and perpendicular to ($I_{\perp}$) the polarized field. The emission anisotropy (A) is then defined as

$$A = (I_{\parallel} - I_{\perp})/(I_{\parallel} + 2I_{\perp}) \qquad 3.42$$

while the decay is related to the relaxation time for the rotation (τ_r) by (compare with eq. 3.6 and subsequent discussion) the expression

$$A(t) = A_0 \exp(-3t/\tau) \tag{3.43}$$

The older method used to measure this property for a macromolecule involved radiating with continuous illumination and measuring the polarization (P, expressed as $(I_\perp - I_\parallel/(I_\perp + I_\parallel))$ when

$$\frac{1}{P} \pm \frac{1}{3} = \left(\frac{1}{P_0} \pm \frac{1}{3}\right)\left(1 + \frac{RT}{V\eta}\tau\right) \tag{3.44}$$

(plus and minus correspond to unpolarized and polarized incident radiation) where P_0 is the polarization that would be observed if no rotation occurred, i.e. the intrinsic polarization, V is the solvated molar volume, and η is the viscosity of the solvent. Thus, in this experiment, there are two unknowns (P_0 and τ) which must be estimated before it is possible to calculate rotational coefficients. To do this, it is necessary to measure P at several values of T/η by changing both the temperature (T) and viscosity of the solvent. Normally, the viscosity is changed by adding inert solutes like sucrose to the solution. Carrying out these additional experiments is tedious, and, obviously, changing both temperature and solvent composition can move conditions away from biological reality, so for these reasons the pulsed procedure is to be preferred.

Because lasers working at 280 or 260 nm (the two usual absorption maxima for proteins and nucleic acids) are not commercially available as yet, workers have resorted to attaching artificial chromophores to proteins and nucleic acids which have absorption maxima in visible light. For example, eosin (2,4,5,7-tetrabromo-fluorescein) was covalently coupled to proteins in the membrane of the erythrocyte, and Cherry *et al.* (1976) 'flashed' this labelled protein *in situ* in order to measure rotational relaxation times of the molecules in the membrane. They found that the rotational diffusion was of the order of $10^3\ s^{-1}$ ($\tau \approx 0.3$ ms) in an axis which was normal to the plane of the membrane. Translational diffusion within the membrane appeared to be extremely slow.

The fluorescent technique can be extended so that the emitted radiation excites a second chromophore in the same molecule. Under these circumstances, the excited second chromophore fluoresces at a different wavelength to the first. By comparing the intensities for the molecule with and without the second chromophore, the efficiency of transfer can be calculated. Transfer can only occur if (*a*) the dipoles of the two chromophores are aligned correctly so that dipole–dipole interactions are favoured and (*b*) there is sufficient overlap between the spectra of the fluorescing first chromophore and absorption of the second. Because the efficiency depends on dipole–dipole interactions, the efficiency decreases according to the r^{-6} law (see Chapter 1) where r is the distance between the chromophores. Despite the many uncertainties in interpreting these data, Rigler and Ehrenberg (1976) were able to use this technique to measure the distance between known points of native t-RNA in solution. In principle, it should be possible to extend this method to measure flexing of the molecule, because flexing will modulate the transfer of energy to the second chromophore with a frequency equal to that of the flexing of the molecule. This has yet to become a routine method, but the work of Rigler and Ehrenberg (1976) suggests that it may be a powerful tool in the future.

3.5 Lessons to be learnt from studying molecular motion

This chapter opened with questions about structure and shape of macromolecules in solution; these questions concerned dynamic aspects of the shape, whether it can be measured using known techniques, and what relevance it has to the biological activity of macromolecules. Having discussed some of the results and techniques, it is possible now to summarize with a few conclusions.

(1) Apart from Raman scattering and other spectroscopic methods, it is not possible to measure structural details in solution. Hydrodynamic methods give information about the overall shape of the macromolecule and some data on its gross movements but not on details of structure.
(2) Molecules do not exist in solution as isolated entities, and the concentration dependence shows that they interact with their own type in predictable but complex ways. Intuitive extrapolation from known thermodynamic activities of small molecules would predict this, but the interesting fact that emergies from hydrodynamic studies is the distance over which these interactions spread. In Chapter 1 is discussed macroscopic effects for particles like TMV produced by balancing van der Waal's forces against repulsive forces, but hydrodynamic studies show that these long-range effects are not confined to large particles but include all macromolecules solvated by water. Long-range interactions are particularly interesting because local concentrations in biological cells can be high, and the macromolecules are never at their thermodynamically ideal state.
(3) Physically based techniques make it possible to study effects caused by changing the concentration of macromolecules. Although the hydrodynamic concentration dependence may produce only second-order corrections when determining overall shapes, the importance of these is large in determining the energy balances between macromolecules and their solvent. The algebraic sum of the interactive energies can give either attractive or repulsive forces between the macromolecules, making it difficult to predict how a concentrated solution will behave.
(4) Because of the effects summarized in (1)–(3), it is important to study, by physical techniques, concentrations of macromolecules similar to those found in their native biological state. Furthermore, since there are marked differences in the interaction of a polyampholyte at low and high concentrations of counter-ions, it is important to work at the physiological concentration of counter-ions. It could be argued that, although it is desirable for physical chemists to include excess counter-ions when studying macromolecules, this is not necessarily the biological environment of the macromolecule.

These four points could be summarized further by saying that, although long-ranging weak forces are included in mathematical models as empirical coefficients, they appear to share with hydrogen bonds and hydrophobic effects the property of determining biological activity of macromolecules.

3.6 Appendix to Chapter 3

3.6.1 Relationship between osmotic pressure and diffusion

In Figure 3.12 is shown diagrammatically the ideal model where the pressure is $A\Pi$ at x (plane 1) and $A(\Pi + \mathrm{d}\Pi)$ at $x + \mathrm{d}x$ (plane 2) (these pressures are positive in the direction right to left in Figure 3.12). Thus, the resulting pressure on a particle in the volume $A\mathrm{d}x$ is $-A\mathrm{d}\Pi$. At any plane, the resultant force (F) per unit volume is $-\mathrm{d}\Pi/\mathrm{d}x$, the osmotic gradient, while for an ideal solution (infinite dilution, see eq. 1.12) the osmotic pressure is

$$\Pi = cRT \tag{3.45}$$

and the gradient is

$$\mathrm{d}\Pi/\mathrm{d}x = RT\,\mathrm{d}c/\mathrm{d}x \tag{3.46}$$

Since the number of solute particles per unit volume is $\mathcal{N}c$, the force acting on each solute particle is

$$F = -\frac{RT}{\mathcal{N}}\frac{1}{c}\frac{\partial c}{\partial x} \tag{3.47}$$

The uniform velocity for each particle (v) in a homogeneous continuous medium and driven by a constant but random force is

$$v = f^{-1}F \tag{3.48}$$

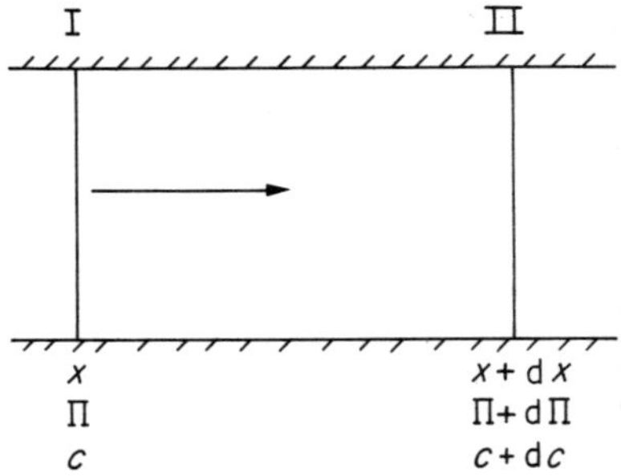

Figure 3.12 Schematic model for osmotic flows through a vessel. Each plane has a surface area A (m^2) with negative flows of solute in the direction of the arrow

Combining eq. 3.48 with eq. 3.47 gives

$$v = -\frac{RT}{\mathcal{N} f}\frac{1}{c}\frac{\partial c}{\partial x} \qquad 3.49$$

Since $v = \mathrm{d}x/\mathrm{d}t$ and $\mathrm{d}x/\mathrm{d}t = \mathrm{d}M/A\mathrm{d}t$, eq. 3.49 can be rewritten as

$$\left(\frac{\partial M}{\partial t}\right)_x = -\frac{RT}{\mathcal{N} f}\left(\frac{\partial c}{\partial x}\right)_t A \qquad 3.50$$

$$\left(\frac{\partial M}{\partial t}\right)_x = -D\left(\frac{\partial c}{\partial x}\right)_t A \qquad 3.51$$

Combining eq. 3.50 with Fick's first law of diffusion (eq. 3.51) shows that

$$D = \frac{RT}{\mathcal{N} f} \qquad 3.52$$

A rigorous derivation can be obtained through the changing chemical potential ($\partial\mu/\partial n$), but the simple derivation given serves to show the functional relationship between osmotic pressure and diffusion.

3.6.2 Frames of reference for kinematic viscosity (diffusion)

The basis of this interpretation is Phillies (1975). In general, theoretical calculations on kinematic viscosity are for cases where the solvent is an immobile continuum, and this is called the solvent-fixed frame (the solvent is denoted by subscript 0). With the exception of isotope experiments, the experimental frame is the scattering or diffusion cell, and this is called the volume-fixed frame of reference (V_i). Thus, for a two-component system in which the partial volumes $\bar{V}_i$ are independent of concentration, we have by definition

$$c_0\bar{V}_0 + c_A\bar{V}_A = 1 \qquad 3.53$$

where c_i are the appropriate ensemble-averaged concentrations expressed in number densities (i.e. $c_A = \langle a(r, t)\rangle$ for component A having a concentration a at point (r, t); the number density equals $\mathcal{N}c/M$ for dilute systems when c is in $\mathrm{g\,l^{-1}}$, $\mathcal{N}$ is Avogadro's number).

Fick's law can be used to define the diffusion coefficient (called a mutual diffusion coefficient by Phillies (1975))

$$-J_A = D_A \nabla a_A(r, t) \qquad 3.54$$

in which J_A is the flux (or current) of species A. Then, applying the conservation of mass to the system, it is known that for the solvent-fixed frame

$$(J_0)_0 = 0 \qquad 3.55$$

Conversely, in the volume-fixed frame, it is known that

$$\bar{V}_0(J_0)_V + (J_A)_V\bar{V}_A = 0 \qquad 3.56$$

From classical kinematics is known the relative velocity of the volume frame to the solvent frame (U) as given by

$$(J_A)_V = (J_A)_0 + c_A U_{V_0} \qquad 3.57$$

(the subscript V_0 refers to the volume frame of the solvent). Pictorially, eq. 3.57 describes the flow of the solvent into spaces left by the Brownian translation of the particles. Multiplying eq. 3.57 by V_i and summing gives

$$(J_0)_V \bar{V}_0 + (J_A)_V \bar{V}_A = (J_A)_0 \bar{V}_A + (J_0)_0 \bar{V}_0 + c_A \bar{V}_A U_{V_0} + c_0 \bar{V}_0 U_{V_0} \qquad 3.58$$

Since eq. 3.56 defines the left-hand side of eq. 3.58 as zero, it is possible to simplify through eq. 3.53 to give

$$-(J_A)_0 \bar{V}_A = -U_{V_0} \qquad 3.59$$

Finally, substituting eq. 3.59 into eq. 3.57 gives the required relationship, i.e.

$$(J_A)_V = (J_A)_0 - c_A \bar{V}_A (J_A)_0 \qquad 3.60$$

which, when substituted into Fick's law (eq. 3.54) gives the relationship for diffusion coefficients, i.e.

$$(D_A)_V = (D_A)_0(1 - c_A \bar{V}_A) \qquad 3.61$$

$c_A \bar{V}_A$ is the volume fraction, Φ, assuming the partial volumes describe the true volumes of the macromolecules. From eq. 3.61 it follows that, since the Einstein equation ($D = kT/f$) is derived for the solvent frame, this equation can only apply when $\Phi = 0$. Even in the absence of interaction between macromolecules, the $(1 - \Phi)$ correction is necessary to convert experimental measurements at finite concentrations into a form suitable for the Einstein equation.

Dynamic viscosity

Here the relationship between experiment and theory is not in doubt, because for a homogeneous fluid carrying particles the inertial terms either disappear or are negligibly small. Thus, for a homogeneous Newtonian fluid of viscosity η_0, the work done by the stresses on the apparatus boundaries is w_0. Suspending particles in the fluid changes the viscosity to η and the work to w, to give

$$\frac{\eta}{\eta_0} = \frac{w}{w_0} \qquad 3.62$$

i.e. one does not need to know about the internal constitution of the fluid to determine viscosity. The Einstein relationship for a sphere acknowledges that the important factor in the concentration dependence of viscosity of fluids is the volume fraction. However, when relating theory to experimental results, often the theoretical derivation uses the ratio of radius to distance between spheres and the resulting coefficients will apply when concentration is expressed as $\Phi^{1/3}$.

CHAPTER 4

Sticky molecules

The experimental results and discussion given in earlier chapters illustrate the ubiquitous sticky property of all proteins. Forces of wide-ranging energies link the molecules in a solution, and, even when covalent forces are excluded from these interactions, there remain the intermolecular forces (discussed in Chapter 1). The strengths of these intermolecular forces vary from weak forces expressed as thermodynamic non-ideality through to forces responsible for stabilizing multisubunit particles. Both convenience and the large differences in energies are the reasons that multisubunit interactions are considered separately from thermodynamic non-ideality. Energies of multisubunit interactions are sufficiently high that the multisubunit has a stable existence in the experimental timescale; in non-ideality, the accretions of molecules are not stable.

Ascribing reasons to why some active macromolecules exist as single molecules while others are more 'sticky' and become aggregates depends on whether the discussion is physico-chemical or biological. Physically, the causes can be discussed in terms of the generation of lyophobic surfaces which interact to exclude water; that is, the explanation lies in changing solvent–solute interactions. But this does not show why these changes are necessary for the biological activity of the macromolecule. Biological explanations try to give causes and involve genetic control of the biological activity through changing construction of the molecules. By implication, this explanation suggests that, because of the unique nature of the construction, lyophobic surfaces become exposed to the solvent (Tanford, 1978). Hence, one explanation invokes the other, suggesting that evolutionary pressures enlist the natural tendency for macromolecules to interact when evolving special biological properties for the aggregate which the individual units are unable to supply. Thus, one finds that some multisubunit complexes exhibit cooperativity between the subunits which could not be generated by long-range interactions involving the solvent; for example, the transport of oxygen by haemoglobin. On the other hand, there exist multisubunit enzymes of similar construction to haemoglobin (i.e. tetrameric) which do not show any cooperativity between the subunits; for example, lactate dehydrogenase. The evolutionary pressures which produced these oligomeric complexes can be seen in some cases to arise from changing physiological and morphological states. Thus, as the organisms became more complex during the evolution and organs were removed further from surfaces exposed to gaseous oxygen, then systems had to be evolved which transported oxygen efficiently. Oxygen is relatively insoluble in water (but soluble in many non-polar liquids). Hence arose the need to evolve a molecule that could complex oxygen and release it under demand at lower partial pressures, and the result was tetrameric haemoglobin. This superficial interpretation may satisfy some questions, but it does not explain why in some organisms the problem was solved through producing very large multisubunit aggregates like haemocyanin in gastropods, while others retained simpler complexes but packaged them at high concentrations in special cells (haemoglobin). It is difficult to support universal arguments for the evolution of oligomeric complexes based on increased efficiency of activity that may accrue from forming the multisubunit enzyme, as for example with lactate dehydrogenase. Here, the requirement to form a multisubunit enzyme must come from unknown pressures acquired when evolving other characters in the organism.

Specific conditions can be arranged which cause any macromolecule to aggregate, and these form the basis of many methods used to separate a mixture into individual macromolecules. As a general rule, conditions which either neutralize net charges or shield the charges by high salt concentrations lead to precipitation of macromolecules; even increasing the macromolecular concentration to high levels may cause precipitation. Although these phenomena are not the subject of this chapter, the interactions involved form a background for discussing the methods of study and relevance of naturally occurring oligomeric complexes.

The fact that many biologically active macromolecules occur natively as complexes suggests that active single chains are rarer than complexes. An idea of the widespread occurrence of multisubunit complexes can be gained from the table produced by Darnall and Klotz (1975) containing over 500 enzymes formed from two or more subunits (these may be of one or more types of subunit). The number of subunits that form one type of active enzyme catalysing a single reaction can vary between the species. In some cases, the changing number of subunits in the complex appears to follow evolutionary development; for example, lactate dehydrogenase is dimeric in crabs and sea-worms (Long and Kaplan, 1973) and tetrameric in mammals (Jaenicke and Knof, 1968). With other enzymes, there appears to be no relationship with the complexity of the organism; for example, phosphoglycerate mutase is isolated as a tetramer from yeast (Darnall and Klotz, 1975) and as a dimer in mammals. Occasionally, the number of subunits remains invariant with species; for example, enolase is dimeric in most species except one thermophyllic bacterium living at 60°C (Darnall and Klotz, 1975). These varying states of complexity must reflect changing interactions between the subunits and solvent produced by the differing amino acid compositions of the chains.

Our knowledge about the 'sticky' nature of macromolecules comes from studying their dissociations in various solutions. One of the first quantitative studies of interacting molecules was made in the 1920s by studying haemoglobin and haemocyanins using the ultracentrifuge (Svedberg and Pederson, 1940) and osmotic pressure (Adair, 1925). However, Gutfreund (1948a, b) was one of the first to obtain results which suggested that the association–dissociation may be more widespread than expected when he demonstrated (using osmotic pressure) the association of subunits of insulin. Since then, both the number of techniques and the understanding of artifacts have increased, so raising the study of association–dissociation of macromolecules to a routine process. Two major monographs have been written on the investigation of these reactions (Cann, 1970; Nichol and Winzor, 1972), and these provide detailed methods for analysing results obtained from transport methods.

4.1 Basic mathematical models of specific macromolecular interactions

In Chapter 2 are discussed problems encountered when measuring relative masses of discrete macromolecules, and in Chapter 3 are discussed the weak interactions which give rise to concentration dependence of hydrodynamic parameters. These discussions are relevant to the present task of studying macromolecular interactions. In particular, supposing tests made for heterogeneity were positive, showing that the sample was paucidisperse, then one must assume either that the purification was incomplete or that specific macromolecular interactions were occurring. Of these two alternatives, only specific interactions are of biological interest, so assuming further attempts to purify the sample did not reduce the heterogeneity, then one must investigate interactions using mathematical models to describe the observations. Discussions of these basic models can be simplified if one restricts the models to those found in the self-association that leads to a dimer (eq. 4.1). In principle, these equations can be expanded to include the case of indefinite polymerization and mixed reactions of the type

$$\mathrm{A} + \mathrm{B} \rightleftharpoons \mathrm{C}$$

(although this last one is much more complex than self-association):

$$2\mathrm{A} \underset{k_{-1}}{\overset{k_1}{\rightleftharpoons}} \mathrm{A}_2 \qquad 4.1\mathrm{a}$$

$$K = [\mathrm{A}_2]/[\mathrm{A}]^2 = k_1/k_{-1} \qquad 4.1\mathrm{b}$$

The relative mass (M_2) of A_2 is twice that (M_1) of A, but if the rates of reaction are fast relative to the time required to separate A from A_2 then no separation occurs and the result will be an average parameter estimated from any experimental results (eq. 4.2). This average parameter can be used for determining K (eq. 4.1b) providing the mixture is at chemical equilibrium. The average parameter can be relative mass ($\bar{M}$), diffusion coefficient ($\bar{D}$), sedimentation coefficient ($\bar{S}$) or elution volume ($\bar{V}$), and the averages could be number- (eq. 4.2a), weight- (eq. 4.2b) or less often Z-average (eq. 4.2c), depending on the experimental procedure. Osmotic pressure gives number-average relative mass ($\bar{M}_\mathrm{n}$, eq. 4.2a), intensity light scattering and gel chromatography give weight-average parameters ($\bar{M}_\mathrm{w}$ and $\bar{V}_\mathrm{w}$), quasi-elastic light scattering gives a Z-average diffusion coefficient ($\bar{D}_Z$), while all three averages can be obtained from equilibrium centrifugation (Roark and Yphantis, 1969; see Section 2.4.1):

$$\bar{M}_{\mathrm{n}} = \frac{\Sigma_{i=1}^{N} n_i M_i}{\Sigma_{i=1}^{N} n_i} = \frac{M_1(c_1 + c_2)}{(c_1 + 0.5c_2)} \quad 4.2a$$

$$\bar{M}_{\mathrm{w}} = \frac{\Sigma_{i=1}^{N} n_i M_i^2}{\Sigma_{i=1}^{N} n_i M_i} = \frac{M_1(c_1 + 2c_2)}{(c_1 + c_2)} \quad 4.2b$$

$$\bar{M}_Z = \frac{\Sigma_{i=1}^{N} n_i M_i^3}{\Sigma_{i=1}^{N} n_i M_i^2} = \frac{M_1(c_1 + 4c_2)}{(c_1 + 2c_2)} \quad 4.2c$$

Generally, it is more convenient to use concentrations (c_i) rather than chemical activities in the equations to give the experimentally determined interaction coefficient (labelled $L_{1,2}$ in the example of dimerization),

$$L_{1,2} = c_2/c_1^2 \quad 4.3$$

Replacing c_2 in eq. 4.3 by $L_{1,2}c_1^2$ and assuming $c_1 + c_2 = c_{\mathrm{T}}$ (total concentration), then the relationship for the average relative masses for an ideal macromolecule becomes

$$\bar{M}_{\mathrm{n}} = M_1(1 + 2L_{1,2}c_1)/(1 + 0.5L_{1,2}c_1) \quad 4.4a$$

$$\bar{M}_{\mathrm{w}} = c_1 M_1(1 + 2L_{1,2}c_1)/c_{\mathrm{T}} \quad 4.4b$$

For most experimental conditions, M_1 (eq. 4.4) must be replaced by $M_1/1 + B_2^{(1)}c_1)$ (confining non-ideality to the second virial coefficient $B_2^{(i)}$, see Chapter 2) in order to introduce thermodynamic non-ideality, so the equation becomes

$$\bar{M}_{\mathrm{n}} = M_1(c_1 + c_2)/[c_1(1 + B_2^{(1)}c_1) + 0.5c_2(1 + B_2^{(2)}c_2)] \quad 4.5a$$

$$M_{\mathrm{w}} = M_1\{[c_1/(1 + B_2^{(1)}c_1)] + [2c_2/(1 + B_2^{(2)}c_2]\}/c_{\mathrm{T}} \quad 4.5b$$

Assuming the molecular shape of A_2 is not significantly different from A, then $B_2^{(1)} = B_2^{(2)} = B_2$, and eq. 4.5 can be reduced to

$$\bar{M}_{\mathrm{n}}/M_1 = c_{\mathrm{T}}/[c_1(1 + B_2c_1 + 0.5L_{1,2}c_1) + 0.5L_{1,2}^2B_2c_1^4] \quad 4.6a$$

$$\bar{M}_{\mathrm{w}}/M_1 = [c_1(1 + 2L_{1,2}c_1) + 3B_2L_{1,2}c_1^3]/[(1 + B_2c_1 + L_{1,2}c_1^2)c_{\mathrm{T}}] \quad 4.6b$$

Similar relationships can be written for diffusion, sedimentation, and elution volumes where the appropriate coefficients and concentration dependence replace relative mass and B_2.

Equation 4.6 forms the basic mathematical model for analysing dimerization between macromolecules. More terms will be required for analysing mixed reactions and indefinite polymerization. These equations form a subset of the generalized multiple equilibrium reactions discussed by Tanford (1961, p. 526). The important idea conveyed by these models is that interactions are governed by the laws of mass action, and values of the average molecular parameters will change with increasing concentration (increasing $S_{20,w}$ and relative mass but decreasing diffusion coefficients). This change is either opposite to that expected from non-ideality or, in the case of diffusion, is enhanced, so that the relationship between the parameter and concentration forms a study which separates interaction from contamination.

4.2 Estimating interaction parameters

4.2.1 Averaged parameters

Using the equations outlined in Section 4.1 for analysing experiments implies that the correct average is calculated from the records. A number-average relative mass can only be calculated directly from osmotic pressure, but by making certain assumptions (see eq. 2.38) this average can be calculated from sedimentation equilibrium. Weight-average relative masses are measured directly from intensity light scattering and centrifugal equilibrium, but these methods do not provide either spatial or temporal resolution of the components in the mixture (in this, they resemble osmotic pressure). Partial resolution can be achieved using transport experiments such as sedimentation velocity, electrophoresis, and gel chromatography, but the profiles will consist of incompletely separated boundaries (as shown in Figure 4.4 and discussed in Section 4.2.2). When analysing these transport experiments, it is common to calculate averaged parameters from each experiment, making it necessary to cover a range of concentrations. These parameters must be calculated from the velocities of the centre of mass of the boundaries and not the peaks of a schlieren record (see Schachman (1959, p. 64) for a discussion of the appropriate equations for the centrifuge and Cann (1970) for electrophoresis). The calculations of these centres are tedious but readily programmed for a computer, and require the measurement of coordinates across the profile from which weighted areas can be calculated. These averaged parameters can then be related to total concentration for each experiment to give a characteristic increase in $S_{20,w}$ (or decrease for diffusion) for interacting molecules. From the form of the dependence can be estimated $L_{i,n}$, but these basic equations do not show which are the crucial variables that determine the final accuracy in the interaction coefficients nor how their errors influence the ability to separate contamination from interaction.

Errors in $L_{1,2}$

The magnitude expected for $L_{i,n}$ must influence both the permissible level of errors that can be tolerated and the technique used to measure the interactions. Accepting these two limits means that a general treatment of sources of errors is not possible, but guides can be given by taking eq. 4.5b as the model. Equation 4.5b contains only three experimental variables for the dimerization ($\bar{M}_w$, M_1, and c_T) and random errors in these will be compounded when calculating $L_{1,2}$. An error analysis of eq. 4.5b produces the relationship shown in eq. 4.7—the variance of $\bar{M}_w$, M_1, and c_T ($= P_i$) is shown as $V(P_i)$. In order to simplify the algebra and produce a compact final expression, it has been assumed that the covariances between the variables are zero and that terms containing c_1/M_1 are effectively zero (that is, assuming $c_1 < 1$ and $M_1 \gg 1$):

$$V(L_{1,2}) = \frac{1}{c_1^4 M_1^2}\left[V(c_1)\left(\frac{\bar{M}_w c_T^2}{c_1^2} - \frac{3\bar{M}_w M_1 c_T}{2c_1} + M_1 + 0.5 \right) + \frac{1}{4}\left(\bar{M}_w^2 V(c_T) + \frac{V(M_1)}{4c_1^2} + V(M_1)c_T^2 \right) \right] \qquad 4.7$$

Order-of-magnitude calculations using eq. 4.7 show that the error in $L_{1,2}$ is determined largely by the first two terms contained in the first large parentheses, and even here it is the error in c_T which is the most important. The conclusion from this simple analysis of errors is that, providing the experiments cover a wide range of concentrations, then even approximate estimates of $\bar{M}_w$ at a series of concentrations should be sufficient to produce acceptable estimates of $L_{1,2}$ as well as to discriminate between an impure sample and one exhibiting specific self-association. Including non-ideality (as in eq. 4.6) obviously improves the accuracy of the estimate of $L_{1,2}$ without appreciably affecting the errors; it increases these by approximately $(1 + B_2 c)$. A similar result would have been found if the dependence of the sedimentation coefficient on interaction (eq. 4.9) had been analysed for errors, but here the main source of error is not random but systematic and caused by uncertainties in the temperature correction used to normalize the measured sedimentation coefficient. Exact absolute temperatures must be measured in order to correct for viscosity and density of the buffers relative to water.

At concentrations less than $1\,\mathrm{g\,l^{-1}}$, neutral (or nearly neutral) globular protein solutions behave as though they were ideal; above this concentration, non-ideality becomes increasingly important, even though this does not appreciably affect random errors. If non-ideality is ignored (or concentration dependence for $\bar{S}_{20,w}$), then the estimates of $L_{1,2}$ will contain systematic errors which will distort the results. At finite concentrations, there will be $(1-\alpha)^n c_T^n$ units of monomer and αc_T units of polymer (α is the degree of association and n is the order of polymerization); changing the concentration induces changes in the relative proportions of the species in the mixture. Each species will have an associated parameter describing its concentration dependence, but, if it is assumed that they all have similar dependence (as was done when converting eq. 4.5 to eq. 4.6), then it is possible to calculate the extent to which non-ideality affects the value of $L_{1,2}$ for dimerization. This effect has been simulated for the change of relative mass caused by the dimerization of a macromolecule having a monomeric relative mass of 66 000 dalton (Figure 4.1). Introducing a positive second virial coefficient of acceptable value into the calculations showed that the $\bar{M}_w$ for the ideal case was only 0.7% greater than a non-ideal spherical molecule even at $20\,\mathrm{g\,l^{-1}}$ (Figure 4.1*b*). Thus, in general, the virial coefficients only become important if relatively high concentrations are examined or the macromolecule is very

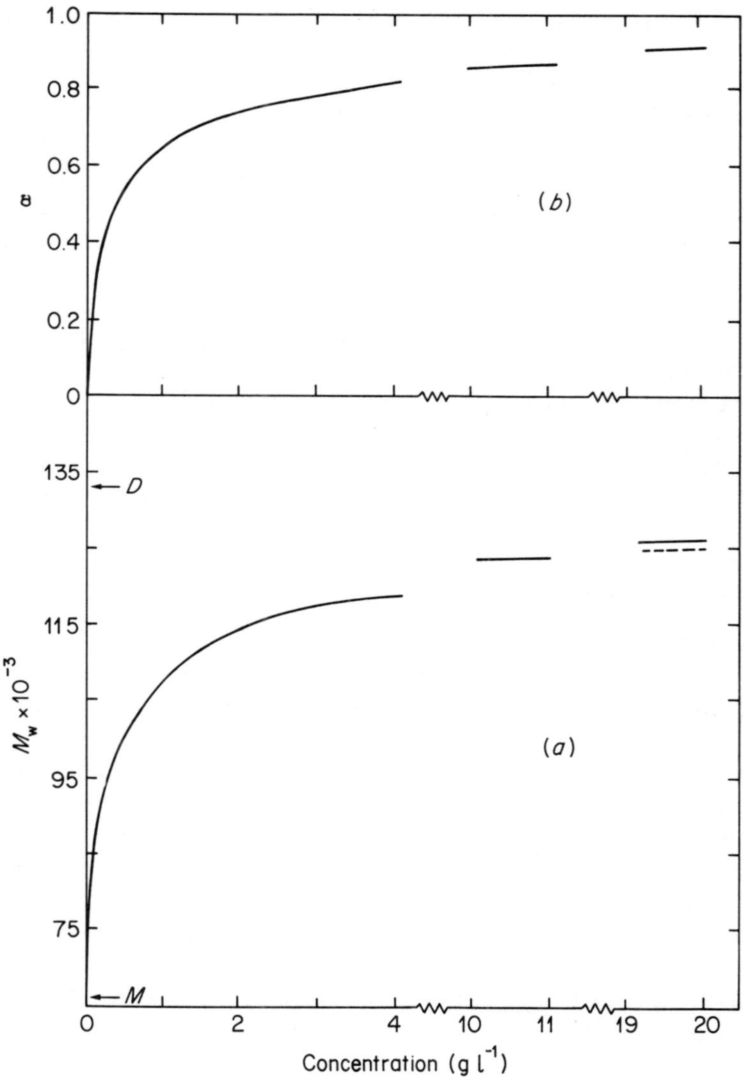

Figure 4.1 The relationship between M_w and concentration for a reversibly associating system. (*a*) M_w for association between a monomer of relative mass 66 000 dalton and $L_{1,2} = 0.5\,\mathrm{l\,g^{-1}}$:ideal (—) and non-ideal (---) results for a $B_2 = 3 \times 10^{-8}\,\mathrm{l\,mol^{-2}\,g^{-2}}$ (the difference between the two plots was not significant at concentrations less than $15\,\mathrm{g\,l^{-1}}$). (*b*) An illustration of the increase in the proportion associated (α) with increasing concentration

asymmetrical or highly charged (when third and higher coefficients may become important). Of course, as the order of the reaction increases beyond dimerization, then the operational value for B_2 increases for the larger molecules and non-ideality becomes increasingly important. As the energy of association weakens, so a point is reached where the increase in $\bar{M}_w$ caused by polymerization is equal to or less than the decrease caused by non-ideality; for the present model, this occurs when $L_{1,2} < 2 \times 10^{-2} \mathrm{l\,g^{-1}}$.

These simulations would not have been so clear had hydrodynamic coefficients been used instead of relative masses. Concentration dependence of $\bar{S}_{20,w}$ is relatively large, and even for a spherical molecule there occurs a 5% decrease in $\bar{S}_{20,w}$ at $10\,\mathrm{g\,l^{-1}}$ compared with the value at infinite dilution. Here, stronger associations than those discussed for the relative masses could be masked by concentration dependence unless the dependences of the individual species are known or could be calculated. If the associations are very weak, then it becomes debatable whether these should be discussed separately from non-ideality or hydrodynamic interactions.

High concentrations of macromolecules can lead to interactions which end with the formation of two phases. One phase contains a weak solution of solute in solvent and the other contains a solution of solvent in macromolecular complex. When this phenomenon occurs, $(\partial \Pi/\partial c)$ goes to infinity and the treatments of interaction given here are too simple to describe the results.

Association in sedimentation velocity

Studies made with the ultracentrifuge dominate the work on interacting systems mainly because the results are readily collected. Analysing these records to give accurate interaction parameters is well described, although the majority of the studies have used the centrifuge in the equilibrium mode rather than in the velocity mode (see Chapter 2 for a discussion of these modes). This is surprising because velocity experiments are simple to perform and give results in a relatively short period. Possibly, more measurements are required from velocity records than from equilibrium (the large number is necessary in order to locate the centre of mass of the boundaries). Furthermore, when using averaged parameters from transport experiments, it is necessary to carry out experiments over a range of concentrations before interaction coefficients can be estimated. This disadvantage may seem to be unique to transport experiments but, as will be seen later, often a range of total concentrations must still be examined with sedimentation equilibrium.

The range of curves which relate $\bar{S}_{20,w}$ to concentration using differing values for $L_{2,4}$ (eq. 4.9) have been calculated by Gilbert and Gilbert (1973), and an example of their simulated curves is given in Figure 4.2. These curves were calculated by including a term for concentration dependence of $S_{20,w}$ (eq. 2.22), and they show that interactions having $L_{1,2}$ as low as $0.5\,\mathrm{g\,l^{-1}}$ could be measured in the centrifuge through $\bar{S}_{20,w}$. The curves show also that, where high associative energies are involved ($L_{2,4} > 100\,\mathrm{g\,l^{-1}}$), the measurements that are useful must be collected at total concentrations less than $1\,\mathrm{g\,l^{-1}}$.

When discussing $\bar{S}_{20,w}$, the associative equation given in eq. 4.5 must be modified to become

$$\bar{S}_{20,w} c_T = c_1 [S_1^\circ(1 - g_1 c_1) + L_{1,2} c_1^2 S_2^\circ(1 - g_2 L_{1,2} c_1^2)] \qquad 4.8$$

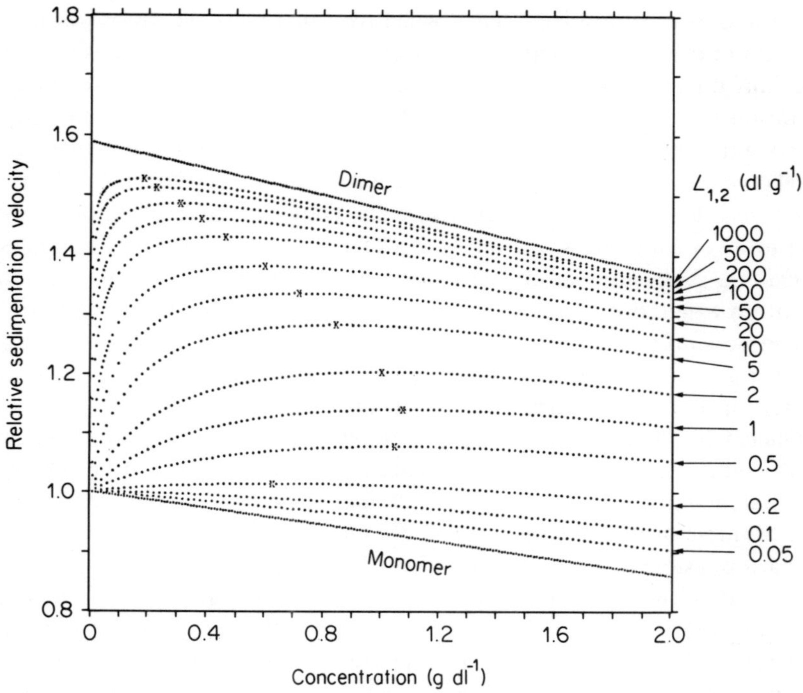

Figure 4.2 Simulated relationship between relative sedimentation velocity and concentration for a range of interaction coefficients ($L_{1,2}$) for dimerization. (From Gilbert and Gilbert, 1973, *Meth. Enzymol.* **27** 273, Figure 1. Reproduced by permission of Academic Press Inc)

where S_1° and S_2° are the $S_{20,\mathrm{w}}$ for monomer and dimer at infinite dilution, while g_i is the concentration dependence parameter (eq. 2.22). Practically, eq. 4.8 can be simplified by assuming that the frictional coefficients of the monomer and dimer are similar (as before; see p. 154), when $S_2 = 2^{2/3}S_1$ and $g_1 = g_2$, to convert eq. 4.8 to

$$\bar{S}_{20,\mathrm{w}}c_{\mathrm{T}} = S_1^\circ[c_1(1-gc_1) + 1.587L_{1,2}c_1^2(1-gL_{1,2}c_1^2)] \qquad 4.9$$

while

$$c_{\mathrm{T}} = c_1 + c_2 = c_1 + L_{1,2}c_1^2 \qquad 4.10$$

this set of equations contain two parameters (S_1° and $L_{1,2}$) which must be estimated from measured values of $\bar{S}_{20,w}$ and c_{T}. Several examples of the application of this type of analysis exist (Gilbert and Gilbert, 1973; Spragg *et al.*, 1977), and the work shows that relatively accurate estimates of $L_{1,2}$ are possible from the ultracentrifuge using this approach.

Separation of components cannot occur during sedimentation of reversibly associating systems if the reaction rates are fast relative to sedimentation. It may appear that the boundary will consist of unresolved components which would give shoulders or unresolved peaks when recorded by schlieren optics, but this will occur only if species greater than a tetramer are produced during association. This limit on the resolution can be seen from a calculation of the maximum separation that will occur if a monomer of $S_1 = 0.3$ ps is mixed with a second component having $S_2 = 0.476$ ps $(S_2 = S_1 2^{2/3})$ and sedimented at 60 000 rev min^{-1} for 8200 s (the time needed for S_2 to move from 6 to 7 cm, the normal usable length of an analytical cell). The boundary centre of S_1 would move to 6.611 cm while S_2 moved to 7 cm if the two species were not reversibly associating. Diffusion would broaden the boundaries so that incomplete separation would occur, but here shoulders or peaks would be recorded. If these two components formed an associating system, then the boundaries would fuse to become one asymmetrical boundary, bearing no particular features, whose shape is determined by the extent of association (as described by the association coefficient). If self-association gave a tetramer, then S_2 would be about 0.756 ps, and the difference between the sedimentation coefficients would be sufficiently large that incompletely separated peaks would be observed with schlieren optics.

An example of the schlieren record obtained for the sedimentation of a self-associating system (canine apolipoprotein A-1) is given by Teng *et al.* (1977) and reproduced in Figure 4.3. The apparent sedimentation coefficient estimated from the sedimentation of the two peaks showed the expected concentration dependence (Figure 4.3), while the apparent weight fractions in the faster boundary increased with increasing concentration of the apolipoprotein. Naturally, the rates of these incompletely separated boundaries cannot be used to

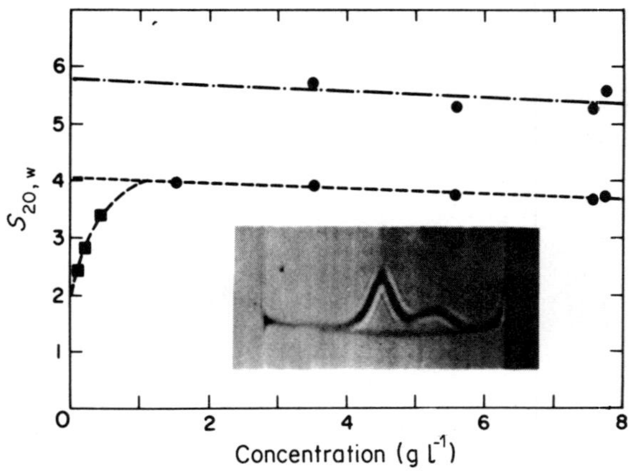

Figure 4.3 The concentration dependence of $S_{20,w}$ (in svedbergs) of canine apo-A-1 lipoprotein in 0.02 M EDTA, 1 mM NaN_3, pH 8.6 at 20°C. The photograph is of the schlieren pattern at a concentration of 5.51 g^{-1} taken 88 min after reaching speed. (Taken from Teng *et al.*, 1977, *J. Biol. Chem.* **252** 8634, Figure 1. Reproduced by permission of the American Society of Biological Chemists, Inc)

determine the individual sedimentation coefficients of the species, nor can the ratios of the apparent concentrations in the peaks be used to calculate an interaction coefficient. Using sedimentation equilibrium, Teng *et al.* (1977) found that apolipoprotein A-1 formed a series of complexes extending up to an octamer, confirming numerically the fact that the $S_{20,w}$ of the faster peak was not that of the octamer (assuming $S_1 = 0.2$ ps as shown in Figure 4.3 for the slow peak, at infinite dilution then S_8 would be approximately 0.8 ps, which is substantially greater than the measured $S_{20,w}$ for the first peak). Useful analyses of the velocities of these boundaries can be made only if the weight-average sedimentation coefficient was estimated for each concentration by using second-moment analysis of each record. Even with incorrectly assessed $\bar{S}_{20,w}$ (Teng *et al.*, 1977, measured the velocities of the peaks), the general shape of the dependence on concentration appears acceptable (Figure 4.3).

An extension of the method based on absolute values of sedimentation coefficients is the differential procedure developed by Smith *et al.* (1973) which is particularly sensitive for associating systems having low energies of interaction. Experimentally, the procedure depends upon recording a differential sedimentation coefficient (ΔS) by employing double-sector cells and interference optics (although similar results can be recorded using absorption optics; McCallum and Spragg, 1972). The reference solution is placed in one sector, preferably cross-linked dimer, etc., and in the other is the equilibrium mixture. Smith *et al.* (1973) developed equations for analysing this system. In principle, this resembles the method developed by Gilbert *et al.* (1973) for gel chromatography, and an example of a typical profile for haemoglobin at low concentration using gel chromatography is given in Figure 2.7.

Association and diffusion

Measurements of diffusion coefficients over a range of concentrations also provide valuable data for determining interaction coefficients of large macromolecules. The nett concentration dependence of diffusion is low and approximately equal to that measured for the dependence of relative molecular mass (see Chapters 2 and 3). However, it should be remembered that doubling the mass of a molecule only reduces the diffusion coefficient by approximately $2^{1/3}$, so sensitivity is reduced compared with the ultracentrifuge. Nevertheless, by using quasi-elastic light scattering to determine diffusion, sufficient accuracy can be obtained to overcome this deficiency and the results from one report on the self-association of myosin made by Herbert and Carlson (1971) show what can be achieved (Figure 4.4). By fitting the appropriate diffusion equation (equivalent to eq. 4.5) to these data, Herbert and Carlson (1971) were able to estimate the equilibrium coefficient plus two second virial coefficients for monomer and dimer. Figure 4.4 illustrates the characteristically large decrease in $\bar{D}$ with small increase in concentration over the critical region where $\alpha \simeq 0.5$ ($< 1\ \mathrm{g\,l^{-1}}$). These particular results were accurate because the large size of myosin makes it a good scatterer, and both types of light scattering (Chapter 3) can be employed to

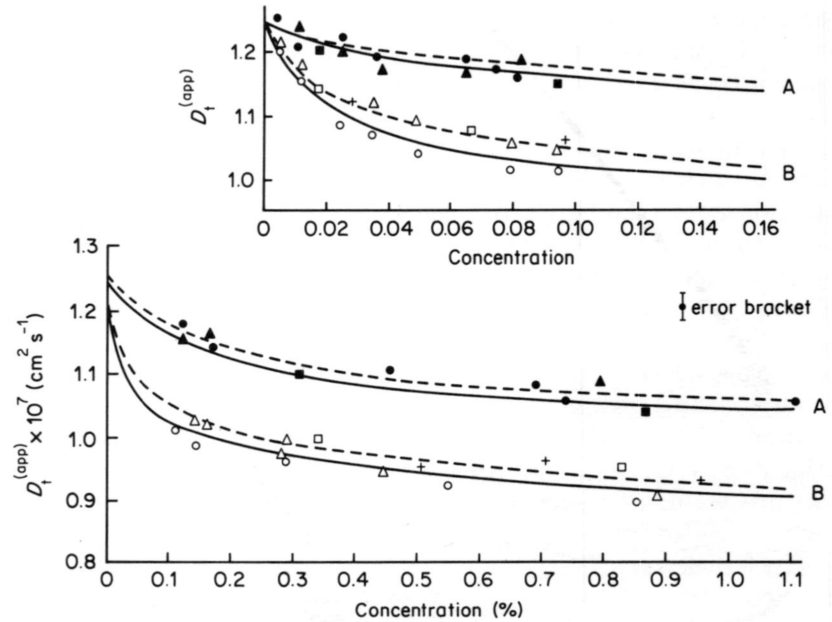

Figure 4.4 Plot of apparent diffusion coefficients of myosin (given in c.g.s. units) as a function of concentration and measured by fluctuation spectroscopy (Chapter 3). The experimental conditions were: homodyne (Δ) and heterodyne (o) spectroscopy in 0.5 M KCL, 0.01 M EDTA, 0.2 M PO_4, pH 7.3; heterodyne (□) in 0.5 M KCL, 0.01 M EDTA, 0.01 M tris, 0.2 M $SO_4{}^{2-}$, pH 7.3; homodyne (Δ) and heterodyne (o) in 0.5 M KCL, 0.01 M EDTA, 0.5 M PO_4, pH 8.5; heterodyne (■) spectroscopy in 0.5 M KCL, 0.01 M EDTA, 0.01 M tris, 0.5 M $SO_4^{2\prime}$ pH 8.5; heterodyne (+) spectroscopy in 2.0 M KCL, 0.01 M EDTA, 0.01 M tris, pH 8.5; The dashed line is for homodyne and the continuous line for heterodyne spectroscopy calculated using the following parameters: $D^{\circ}_{\text{monomer}} = 1.239 \times 10^{-11}\,\text{m}^2\,\text{s}^{-1}$, $D^{\circ}_{\text{dimer}} = 0.838 \times 10^{-11}\,\text{m}^2\,\text{s}^{-1}$, $(B_2 - B_2')_{\text{monomer}} = 0.00561\,\text{l g}^{-1}$ (eq. 3.10b), $(B_2 - B_2')_{\text{dimer}} = 0.00161\,\text{l g}^{-1}$. Two interaction coefficients were tried: A, $L_{1,2} = 0.131\,\text{l g}^{-1}$ and B, $L_{1,2} = 1.061\,\text{l g}^{-1}$. (Taken from Herbert and Carlson, 1971, Figure 2)

measure interaction of large macromolecules at low concentrations where sedimentation is not reliable (the relative mass of myosin is about 450 000 dalton). Results from light-scattering studies of interaction of smaller molecules, of sizes comparable with haemoglobin, are less reliable at low concentrations. If self-association for this size of molecule is not measurable at concentrations suitable for study by ultracentrifugation, then absolute measurements of interaction coefficients are generally not possible.

4.2.2 Boundary analysis of associating systems

Employing averaged parameters to study interacting systems destroys much of the information given by the transport experiments. Specific interactions between macromolecular species distorts the boundaries in precise ways, and the shapes that these boundaries assume during association can be described by integral solutions of the correct differential equations. If the rates of chemical reaction are fast relative to the rate of transport, then the transport of each species will be correlated, making it impossible to separate the species. The result is a distorted boundary (see Figure 4.3) whose shape will depend on the total concentration of the macromolecule. The shape of the profile will change from one which favours the monomer at low concentrations to one containing mainly polymer at high concentrations (it is for this reason that centre of mass or concentration must be calculated when estimating averaged parameters from transport experiments). Conversely, if the rates of chemical reactions are slow relative to the velocities of the boundary, then separation between components is possible, but the separated fractions would re-equilibrate to reform their equilibrium composition if they are isolated and allowed to stand. Re-running the separated fractions after standing would reproduce the original multiple boundaries but of different relative proportions between the features because the total concentrations would

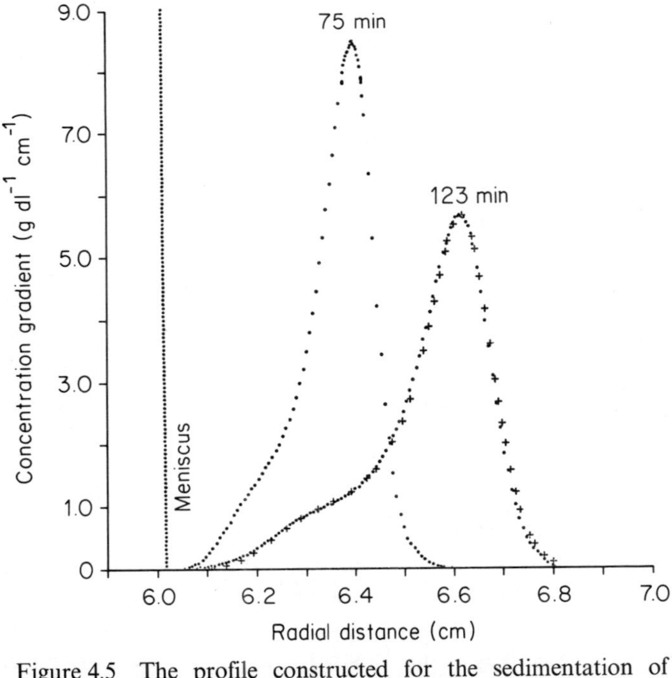

Figure 4.5 The profile constructed for the sedimentation of β-lactoglobulin A under associating conditions (15 g l^{-1}, 25° C, sodium acetate buffer pH 4.65 and ionic strength 0.1 M): . . . . . , measurements made after 75 and 123 min of sedimentation by Armstrong and McKenzie (quoted by Gilbert and Gilbert, 1973); + + + +, computed schlieren pattern for 123 min normalized on the pattern after 75 min, using the method of Cox (1971). (Taken from Gilbert and Gilbert, 1973, *Meth. Enzymol.* **27** 273, Figure 5. Reproduced by permission of Academic Press Inc)

be different. This can be confusing when purifying macromolecules, but if association is suspected it is often worth re-examining a 'pure' fraction after it has stood for a day or so.

Few studies have been reported where simulated profiles have been compared directly with experimental results. It is possible to estimate interaction parameters from such a comparison by iterating between comparison and recalculation after changing the values of the parameters. The calculations are complex, involving numerical integration of the Lamm equation (eq. 2.14) for sedimentation, in which the parameters are not only dependent on concentration but also the equation contains S and D for more than one functionally related species: the relationship between the species is through the interaction coefficient. Nevertheless, the iterative procedure is capable of yielding accurate estimates with data from one experiment, providing the experimental coordinates are measured accurately. Gilbert and Gilbert (1973) have made progress in applying this method by using the numerical methods developed by Cox (1971) and comparing the simulated with experimental results through a computer display. An example of the close agreement that they were able to achieve between calculated and measured profiles is shown in Figure 4.5. The estimated $L_{1,2}$ given by this comparison agreed well with accepted results for β-lactoglobulin (see Table 4.1 for comparison with the $L_{1,2}$ given in the legend of Figure 4.5). This method is potentially one of the most sensitive methods available for studying interaction, but requires considerable expertise to produce reliable estimates.

4.2.3 Equilibrium centrifugation

Applying the centrifuge in this mode makes it possible to study association–dissociation at both true chemical and centrifuge equilibrium, and the only concentration dependence of the parameters arises from thermodynamic and not hydrodynamic non-ideality. Unfortunately, the method is not as sensitive as transport methods for determining the stoichiometry of the reactions. Several mathematical models have been proposed by different workers for analysing these results, and an authoritative survey of these is given by Kim *et al.* (1977). In order to appreciate the method, it is necessary to realize that the records from an experiment involving interacting species yield a non-linear relationship between $\ln c_T$ and r^2 (eq. 2.15b). Thus, results from one experiment would be indistinguishable from results obtained for a single charged species showing large non-ideality, although it is fair to say that this situation has never been reported. This means that the first requirement is to demonstrate that reversible association between the macromolecules is occurring by testing for purity through other procedures (e.g. gel electrophoresis under denaturing conditions to demonstrate purity of the sample). Assuming reversible association is occurring, then the records can be analysed by one of the several procedures given by Kim *et al.* (1977).

One general method is based on first estimating weight-average relative masses at several radial positions by calculating derivatives of the $\ln c_T$ versus r^2 relationship. This gives the relationship between $\bar{M}_w$ and concentration shown in

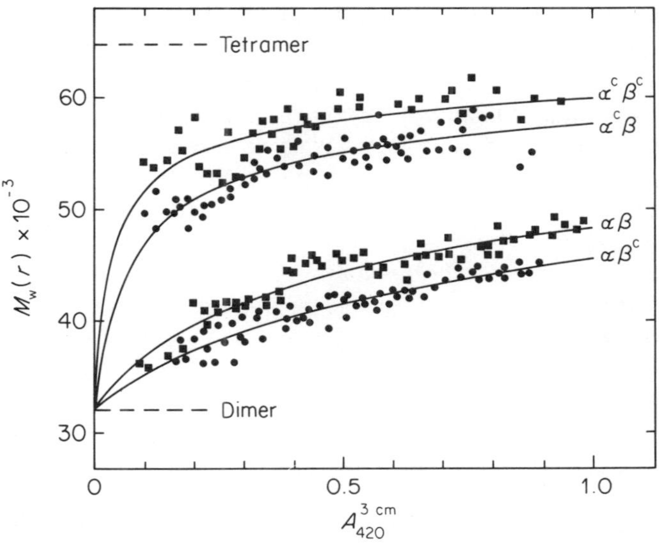

Figure 4.6 $\bar{M}_W$ for normal and carbomylated (c) carboxyhaemoglobin in 0.1 M NaCl, 0.05 M tris, 1 mM EDTA, pH 7, ionic strength 0.15 M. The continuous lines were fitted by least-squares analysis. The results were collected by centrifugal equilibrium at 20°C, 20 000 rev min^{-1} after 20–24 h of running. (Taken from Williams and Kim, 1976, Figure 1. Reprinted with permission from *Biochemistry* **15** 2207. Copyright by the American Chemical Society)

Figure 4.6. The point derivatives must be evaluated numerically, making it important first to smooth the experimental record by fitting an empirical polynomial and using the derivative of this polynomial to calculate the $\bar{M}_w$ within the range of the record. A full analysis of this method, with examples and discussions of propagation of errors, is given by Kim *et al.* (1977). Other methods depend on relating absolute total concentration at each radius to molecular masses through integral equations similar to eqs. 2.24 and 2.25. Examples of the application of this method are given by Sophianopoulis and van Holde (1964), Haschmeyer and Bowers (1970, Reinhardt and Squire (1965), Kellett (1971), Nichol *et al.* (1976) and Spragg *et al.* (1977).

There are lower limits of concentration which can be studied accurately with these methods, and the limits are determined mainly by the sensitivity of the recording procedure. Interferometric methods yield accurate results down to about 0.5 g l^{-1}, while light absorbance generally can extend down to about 0.1 g l^{-1} depending on the extinction coefficient of the macromolecule. At these lower limits, it is difficult to measure accurately the concentrations or their gradients for the technical reasons discussed in Chapter 2. One of the most-studied associating systems is the dimer–tetramer reaction of oxyhaemoglobin; three good reasons account for this popularity. First, the energetics of the dissociation are important to an understanding of the cooperativity found to occur during oxygen transport; secondly, this dissociation reaches 50% at about 0.4 $g l^{-1}$; and, thirdly, the extinction coefficient of the haem–protein is high at 410nm. The last two experimental reasons point to a common problem in studying association, and even with haemoglobin, where its high light absorption makes it possible to work at comparatively low concentrations, the resulting errors are still large. For example, Williams and Kim (1976) found it necessary to work at concentrations around 0.03 $g l^{-1}$ when studying the association of carbamoylated haemoglobins A and S in order to approach a reasonable level of dissociation (the absorbance of the solution given in Figure 4.6 was about 0.25). Despite the apparently high absorbance for this experiment, the spread in their recorded data can be assessed from Figure 4.6. For these molecules, even at the lowest usable levels (about 0.004 $g l^{-1}$), the percentage of associated molecules was considerably greater than 50%. Rarely does one find macromolecules which have sufficiently high absorbance that records can be gathered even at this concentration, and yet many associative energies are sufficiently high that it is desirable to work at concentrations lower than 0.004 $g l^{-1}$. These technical problems limit the ranges of interactions that can be studied with the ultracentrifuge to medium- and low-energy interactions. Instrumental problems do not limit the study of high concentrations; here it is the necessity to include thermodynamic non-ideality in the analysis which increases the complexity of the work and limits the final accuracy of the estimates.

4.2.4 Qualitative tests for association

In considering the theory and examples given in Sections 4.1 and 4.2, it has been assumed that the stoichiometry of the associative reaction is known. For example, with haemoglobin (Figure 4.6) it is known that the tetramer dissociates to give two stable dimers; for myosin (Figure 4.4) it is assumed that the association produces dimers only. If the order of the reaction has to be determined as well as the other parameters (interaction coefficients, concentration dependence of the parameters, etc.), then normally it is not possible to arrive at a unique solution when experimental records are perturbed both by random and systematic errors. Many cases exist where verbal arguments over the correct stoichiometry of an associative reaction have been bitter and lengthy. Fortunately, this need not be repeated in the future because there now exist methods which can usually resolve the issue. Possibly the simplest method involves cross-linking the subunits with bifunctional reagents, such as *bis*-dithio(succinimidopropionate) (Lomant and Fairbanks, 1976). With this procedure, the concentrations of protein must be adjusted to a level which will give both free and complexed subunits in solution because reaction with the reagent may prevent dissociation (however, see below). After the reaction, the components in the mixture are separated by sodium dodecylsulphate gel electrophoresis and the number of discrete species can be compard with electrophoresis of the unreacted control. This method is valid as long as the rates of intersubunit and intrachain cross-linking is greater than either the forward or backward rates of the associative reaction or rates of solvolysis of the reagent. Normally, these conditions occur because, if relatively high molar concentrations of the bifunctional reagent are used, the rates of cross-linking are determined simply by chemical rates (greater than $10^{-3}\,s^{-1}$), while the associative rates in dilute solutions are diffusion-controlled (rates greater than $10^{-2}\,s^{-1}$). Complete intersubunit cross-linking at high concentrations is unlikely to occur because solvolysis of the reagent competes with the cross-linking reactions; thus, raising the concentration of the reagent increases the proportional loss by solvolysis. The other factor which at first sight could appear to confuse the issue is the possibility that cross-linking between the free chains present in high concentrations could give results resembling in every way intersubunit cross-linking. This is unlikely to occur because the stable distance between free molecules in solution (see Chapter 3) is greater than the distance between the two functional groups of the reagent. Observations made by the author suggest that these ideas are borne out by experiments, since it was found that oxyhaemoglobin did not produce polymers greater than the expected tetrameric form until the concentrations were in excess of $50\,g\,l^{-1}$ when using *bis*-(dimethylsuberimidate) as the bifunctional reagent (pH 8.5). The other interesting, but as yet unexplained, observation from these experiments was that random cross-linking with species other than haemoglobin was unlikely to occur in a mixture: the cross-linking appears to be selective for like species. This selection has been reported by other workers. For example, Birnbaumer *et al.* (1979) showed that only adjacent subunits of progesterone–receptor subunits in a cytosol made from oviducts were cross-liked. These observations were made using progesterone labelled with ^{14}C, and the separations made after cross-linking suggested that the concentration of random polymers containing ^{14}C was insignificant. These considerations suggest that this method has great potential for studying associations (as with lactate hydrogenase) where very dilute solutions are required to induce significant amounts of dissociation.

Thomas and Butler (1977) used cross-linking with dimethylsuberimidate to confirm their ultracentrifugal results when determining the limiting maximum size of fractions of the histones. These results showed a predominance of octameric histone, with virtually no lower complexes present. The ultracentrifugal results gave similar findings, but the interpretation of these quantitative results involved fitting statistical models to the data, whereas results from gel patterns were visually obvious. With some reagents, cross-linking has a serious practical disadvantage: many reagents will only react with unprotonated amino groups, so reactions only occur at pH > 8.3 (that is, between adjacent lysine groups); other reagents react with sulphydryl groups, so these work over the normal range of pH providing there are free groupings available on the chain. It is interesting to find that few reagents appear to have been described which are suitable for cross-linking carboxyl groups.

Immunological tests are often employed as definitive criteria of purity. Here antibodies prepared from purified fractions are diffused separately from the antigen through a gel to meet and produce visible precipitates; the number of visible lines gives an indication of the purity of the sample. This test would produce unambiguous results of purity in the presence of self-association providing relatively fast reactions were occurring (remembering that diffusion takes several hours to complete). Under these conditions, a single band would be produced whose shape may be distorted by the association, but this is unimportant because generally the tests are qualitative and only record distances moved but not the shape of the band. However, with slow reactions, the result would indicate heterogeneity unless only one of the polymeric species reacted with the antibody.

4.3 Association of biological macromolecules

The equation describing dimerization (eq. 4.5) contains the following variables: $\bar{M}_w$ or $\bar{M}_n$, total concentration c_T, and M_1 or M_2. If the stoichiometry of the reaction is unknown, then this also must be determined (see Section 4.2.3). Of these variables, precision in measuring total concentrations (see eq. 4.7) determines the accuracy of $L_{1,2}$. Maximum sensitivity in measuring association occurs at the total concentration where equal proportions of the species occurs. The extent of association for biological macromolecules covers a wide range of energies, leading to a corresponding range of concentrations, so no single technique can be universally applied. In order to discuss the problems encountered at each level, it is convenient to divide the range into energetically strong, medium, and weak interactions, as determined by the magnitudes of the interaction coefficients relative to $\Delta G/RT$ (eq. 1.27). It is convenient also to use dimerization to illustrate the discussion, since doubling the reciprocal of $L_{1,2}$ gives the concentration where 50% association–dissociation occurs. Since this will be the point where the maximum effects of interaction can be recorded, it is possible to relate magnitudes of interaction coefficients to usable concentrations. Even at moderate values for $L_{1,2}$ ($> 10^2\,\mathrm{l\,g^{-1}}$), the concentration for 50% association is less than or equal to $2\,\mathrm{mg\,l^{-1}}$, so these are classified as strong interactions. While at the other extreme, where $L_{1,2} < 1\,\mathrm{l\,g^{-1}}$, the concentration will be greater than $100\,\mathrm{g\,l^{-1}}$, and these interactions are weak. It is within the range limited by these two levels that most reported values for biological macromolecules lie. This does not mean that all macromolecular interactions of biological importance fall within this range, but it reflects the limits of the techniques used to measure interaction.

4.3.1 Strong interactions

Darnall and Klotz (1975) list about 500 active macromolecules containing two or more subunits and, of these, less than 5% have been shown to dissociate reversibly upon dilution. Thus, most multisubunit proteins come within the category of strong interactions ($\Delta G/RT > 16$), and it is a salutary lesson to find that no satisfactory accurate physical technique is available for measuring these strong interactive energies without the addition of modifying chemicals. Measurements of coefficients for strongly interacting systems are difficult because either they can only be induced to dissociate under non-physiological conditions or significant quantities of the subunits can only be produced at extremely low dilution. Normally, dissociating reagents act only when present in relatively high concentrations, and this produces its own problems when interpreting hydrodynamic records: these are illustrated by following the studies on lactate dehydrogenase (LDH) in solution. Even if physical techniques were available for studying dissociation of complexes at very low dilutions without the need to include dissociating agents, the inherent instability of the separated subunits can fog the issue from experiments. Here, the results on the dissociation of dimers of oxyhaemoglobin are used to illustrate the problem. These two tetrameric complexes are in themselves contrasting types, in that considerable evidence exists to show that each subunit of LDH binds ligands without cooperativity between the subunits, while tetrameric haemoglobin binds oxygen with cooperativity. Normally, only one type of subunit is incorporated into a tetrameric complex of LDH, but two isozymes exist which differ in the type of subunit complexed. This contrasts with haemoglobin, where two different subunits (α and β) associate to form the tetramer (α_2, β_2). Measuring the binding energies of these two contrasting tetramers could provide quantitative evidence on the biological advantages gained in evolving tetrameric molecules in which the subunits appear to act autonomously in one case but not in the other.

Lactate dehydrogenase

Early measurements of the relative mass of LDH ranged between 115 000 and 154 000 dalton. The uncertainty caused by having such a wide range provoked discussions on whether the molecule was a trimer or tetramer and, if it was a tetramer, did it dissociate upon dilution. Eventually, it was shown that LDH contained only one type of subunit having a relative mass of between 30 000 and 36 000 dalton. Possibly the strongest evidence that LDH dissociated upon dilution came from the work of Miller (1962). He showed that the $S^{\circ}_{20,w}$ was 0.76ps in 0.2 M phosphate buffer, pH 6.0, and about 0.72 ps in 1 M NaCl: this small decrease produced by high salt concentration would be consistent with the molecule dissociating. This view was supported further when it was found that $S^{\circ}_{20,w}$ continued to decrease when higher concentrations of NaCl were used. These observations (plus the later ones of Mire, 1969) were made at a concentration of protein which was less than 0.5 g l^{-1}, but even at this level the dependence of $S_{20,w}$ on concentration was linear and had a negative slope, suggesting that LDH did not dissociate upon dilution. Thus, the evidence for dissociation rested solely on the observation that adding excess salt could reduce $S^{\circ}_{20,w}$. Mire's work (1969) extended the concentration of LDH down to 30 mg l^{-1} without salt, and yet $S_{20,w}$ still remained constant at about 0.75 ps. These experiments required the application of a special technique because at low concentrations the boundaries are not only relatively unstable in the absence of an inert density gradient (Fuller Noel and Schumaker, 1973, and Figure 2.2) but they are also too low to be reliably detected by conventional optical methods. The technique used was the active sedimentation procedure of Cohen and Hahn (1965) in which a zone of the enzyme is layered on a sucrose-gradient-containing substrate which when converted changes its spectrum (e.g. NAD/NADH). Hence, as the zone travels along the gradient, its presence can be detected from the change in light absorption. Conditions can be arranged such that the substrate at any radial plane is not completely converted during the time required for the zone to move through the plane. In other words, the absorbance is a true measure of active enzyme concentration at all times. This can be tested by performing two experiments, one in which lactate is converted to pyruvate (NAD^+ to NADH) and the other in which is measured the reverse reaction, i.e. pyruvate to lactate (NADH to NAD^+), and the two $S_{20,w}$ should agree: this Mire showed to hold for his experiments. A further cause of failure in this type of experiment is where only one of the components (e.g. tetramer) is active. Then, one would measure only the sedimentation of this active component. The experiment therefore requires the additional check to determine that the concentration of the active boundary as well as its $S_{20,w}$ are as expected. The drop in concentration as compared with the static concentration would be greater than that expected by radial dilution alone if the complex was reversibly dissociating and only one form was active. This precaution was not followed with LDH, so there still remained some ambiguity in the interpretation of Mire's results, since it is possible that only tetrameric LDH is active.

This series of experiments showed that for LDH there was an apparent paradox; the enzyme could not be induced to dissociate by dilution, yet comparatively mild treatments with salt caused dissociation. Fortunately, by the time Mire (1969) had carried out the work, an accurately determined relative mass of the tetramer was available (142 000 ± 3600 dalton, Jaenicke and Knof, 1968). This value was calculated from combined measurements obtained with intensity light scattering, ultracentrifugal velocity and equilibrium, diffusion, and osmotic pressure, and it was significant that there were no differences between $\bar{M}_n$, $\bar{M}_w$ and $\bar{M}_Z$. Thus, no dissociation could have occurred under the conditions used in these experiments. Mire (1969) was able to explain the paradox by including the effect that bound salt has on $S_{20,w}$ (Cox and Schumaker, 1961, using earlier work developed by Katz and Schachman, 1955). Cox and Schumaker (1961) showed that, if salt and water bind to a protein, then the relative mass of the hydrodynamic unit (M_h) is given by

$$M_h(1 - \bar{V}_h\rho) = M_p(1 - \bar{V}_p\rho) + \alpha M_w(1 - \bar{V}_w\rho) \qquad 4.11$$

(subscripts p and w refer to protein and water) where ρ is the density of the medium, $\bar{V}_{h,p,w}$ are true partial specific volumes (Eisenberg, 1976), and α is the proportion of preferentially bound water (moles of H_2O per mole of protein, normally unknown). For practical purposes, it is obvious that if $M_h(1 - \bar{V}_h\rho)$ varies non-linearly with ρ in different salt solutions, it is impossible to separate the variables in eq. 4.11, since the amounts of both salt and bound water may change with conditions. If, however, the relationship is linear, then α can be determined from the plot. Equation 4.11 can be rewritten using sedimentation coefficients to give

$$(S\eta)\mathcal{N}f/\eta = M_p + w_w\alpha - (M_p\bar{V}_p + \bar{M}_w\alpha\bar{V}_w)\rho \qquad 4.12$$

where η is the viscosity of the solution. Again, the relationship between $(S\eta)$ and ρ for a variety of salts should be linear if simple binding of the salt occurs. Cox and Schumaker (1961) showed, for a variety of salts, that eq. 4.12 described the properties of ribonuclease and bovine serum albumin. With this foreknowledge, Mire (1969) tested the equations on results for differing salt concentrations containing LDH and found this plot to be linear also, making it possible to calculate the hydration (≈ 0.2 g H_2O/g protein), and the results showed that salts were bound to the protein. This binding caused the decreased sedimentation rates of LDH in high salt by changing either the apparent partial specific volume of the protein or the frictional coefficient (f, eq. 4.12). The preferential hydration appeared to be constant over all salt concentrations employed, although compared with estimates from other proteins (p. 107) the value seems low.

The history of this work demonstrates the importance of treating the solution as a whole system instead of only considering the macromolecule as an inert unit with respect to the solvent. Even so, the apparently successful outcome is still questioned because Cho and Swaisgood (1973) found that storing LDH in concentrations less than 4 g l^{-1} at 20°C, pH 7, resulted in dissociation; this could be prevented if either NADH or dithiothreitol was present in solution. It is still possible that earlier reports of dissociation were the result of this phenomenon. The slow dissociation would not have been detected in fresh solutions. Dissociation of LDH can be induced by adding ethylene glycol to the solution (Miller, 1974). The interpretation of the experimental results as being caused by dissociation could also be explained by arguments similar to those used for NaCl because the glycol appears to bind to the protein. This criticism can be discounted since in experiments with glycol the estimated relative mass was markedly reduced and showed the form of concentration dependence expected for a dissociating system. Furthermore, the partial specific volume of glycol is not sufficiently different from water that the changes in $\bar{V}$ of the complex become important (see Timasheff and Inoue, 1968; Reisler and Eisenberg, 1969; Lee and Timasheff, 1974). This particular property of alcohols makes them attractive reagents for inducing dissociation of oligomeric proteins when using either light scattering (Inoue and Timasheff, 1968) or ultracentrifugal equilibrium experiments. They are less satisfactory for ultracentrifugal velocity experiments because the production of enhanced baselines in the schlieren profiles, which are caused by sedimentation of the glycols, introduces extra uncertainties into the calculations. Detergents are even less satisfactory for hydrodynamic measurements because not only do they bind strongly to protein, and are often charged, but they also exist in solution as an equilibrium mixture between micelles which sediment independently of the complexed ligand, again producing enhanced baselines.

The final answer on the dissociation of LDH came from hybridization studies with heart LDH (H_4) and skeletal muscle LDH (M_4) (these two isozymes have small differences in amino acid composition). Dilute solutions of M_4 and H_4 require many minutes before detectable hybrids such as M_2H_2 form (Miller *et al.*, 1971). If any dissociation of LDH had occurred, then the time to reach significant hybridization would have been measured in milliseconds since it would be limited only by diffusion of the subunits. Furthermore, if dissociation had been proven to occur using hydrodynamic methods but not in hybridization, then the inevitable conclusion would have been that an energy barrier existed for the reaction

$$M_4 + H_4 \rightarrow 2M_2H_2$$

which was not present in the reactions

$$M_4 \rightleftharpoons 2M_2 \qquad \text{or} \qquad H_4 \rightleftharpoons 2H_2$$

This infers that configurational changes occur during the formation of the mixed tetramer which are not present for homologous tetramers, and this is unlikely in two such similar subunits.

Haemoglobin

The dissociation of tetrameric oxyhaemoglobin into two dimers (α, β) is discussed later in this chapter (p. 176) where this reaction is classified as a medium-strength interaction. Both tetrameric deoxyhaemoglobin and the subunits forming the dimers show strong interactions, although for a period there was doubt whether the dimers were stable under normal experimental conditions. Oxyhaemoglobin dissociates to give $2(\alpha\beta)$ with $L_{2,4}$ of about $20\,\mathrm{l\,g^{-1}}$ at 20°C (Table 4.2), and Schachman and Edelstein (1966) found that in very dilute solutions ($< 10\,\mathrm{mg\,l^{-1}}$) monomers could be detected by ultracentrifugal experiments. This result contrasted with earlier results of Ackers and Thompson (1965) and Chiancone *et al.* (1968), where no monomers could be detected. Both these sets of investigators employed gel chromatographic techniques to study the protein, using boundary analysis in the first case and differential procedures in the second (see Chapter 2). The discrepancy between results from the two techniques was resolved by Kellett (1971) who performed experiments in the ultracentrifuge at low concentrations in solvents containing ethene diamine tetraacetic acid (EDTA). When EDTA was present, monomers were absent, showing that chelation stabilized the solution and inhibited denaturation at low concentrations. Further confirmation that the dimer was stable was obtained by a non-physical technique, complement fixation experiments made by Bucci (1971). Bucci tested for hybrids between rabbit and human oxyhaemoglobin at concentrations of $5\,\mathrm{mg\,l^{-1}}$, and showed that no hybridization between dimers was found to occur.

Large changes in interaction energies between subunits of haemoglobin upon deoxygenation were predicted by X-ray crystallography, where Perutz (1970) and Fermi (1975) showed that deoxygenation was accompanied by the formation of two new salt bridges between the subunits. The amino group of lysine $40\alpha_1$ links with the carboxyl of histidine $146\beta_2$, and that of lysine $127\alpha_1$ links with the carboxyl of arginine $141\alpha_2$. Chemically, the effect of these new linkages is to raise the pK of the amino groups. In liganded haemoglobin, only one lysine can form an internal salt bridge in the α-subunit (between lysine 127 and the carboxyl of aspartic acid 6α). It was comforting, therefore, to find these predictions from X-ray diffraction confirmed by Thomas and Edelstein (1972) and Ip *et al.* (1976), who showed that the four subunits of deoxyhaemoglobin were strongly bound together. Forming new salt bridges raises the associative energy between the dimers by orders of magnitude, making detection of dissociation by physical procedure virtually impossible.

The incorrect conclusions drawn from results with LDH in high salt were repeated at one stage with haemoglobin, producing the so-called 'salt paradox'. Basically, this stemmed from results obtained on the oxygenation of haemoglobin, when it was found that cooperativity occurred between the subunits when the protein was dissolved in 2 M NaCl and at concentrations of haemoglobin of about 250 $\mathrm{g\,l^{-1}}$. At this concentration, association should almost be complete, yet the values obtained for $S_{20,\mathrm{w}}$ in this solvent and at this concentration were lower than expected, suggesting that partial dissociation of the tetramer had occurred. Edelstein and Gibson (1971) resolved the paradox by showing that the extent of dissociation calculated from $S_{20,\mathrm{w}}$ had been overestimated because $\bar{V}$ and not ϕ' (eq. 2.20) had been used to calculate the corrected coefficient (this work is discussed in detail by Eisenberg (1976), but the approach is similar to that described for LDH, p. 170).

It would be possible to continue with further detailed discussions of examples of particles possessing strong non-covalent links between subunits and which do not dissociate at usable concentrations. For example, the interaction between nucleic acids and proteins to form virus is sufficiently strong that even single particles do not dissociate *in vitro*, making it necessary to add dissociating agents to separate the two macromolecular types. In this example, the associative energies exceed thermal energy by orders of magnitude, making it impossible for the subunits to move sufficiently far from their neighbours that dissociation can occur. This energy may arise from a combination of hydrophobic energy (entropic) and coulombic energy (enthalpic) to give large interactive energies for association.

4.3.2 Medium interactions

A limited number of macromolecules fall within this category but, because their interaction constants fall within the range for $\Delta G/RT$ of 4 to 16, they can be measured by one of the classical hydrodynamic techniques (see Chapters 2 and 3). Thus, a considerable literature has accumulated on the physico-chemical characteristics of the reactions. Some examples of associating systems that have been studied together with their interaction coefficients are given in Table 4.1. This is a limited list including only cases where values of the coefficients have been calculated. If the list had included qualitative observations as well as measurements made in denaturing reagents, it would have been longer. It would seem

Table 4.1 Selected association constants for macromolecular systems. These are not intended to be definitive values, but are examples of the range of values and orders of interactions found for proteins

Protein	Monomer molecular mass (dalton)	Method[a]	Conditions	$L_{i,n}$(l g^{-1})[b]	i,n	References
β-lactoglobulin	18 336	SE	pH 2.64, $I = 0.16$:			
			25°C	2.3	1,2	1
			5°C	5.04	1,2	1
β-lactoglobulin	18 336	LS	pH 2–3, $I = 0.1$, 25°C	1.5	1,2	2
Glutamic dehydrogenase	3.2×10^5	SE	pH 7.0	2.1		3
		LS	pH 7.0	2.1		3
		QLS	pH 7.0	2.25		3
Lysozyme	13 000	SE	pH 8	0.034	1,2	4
Tyrocidines	1900	SE	30% ethanoic acid, 0.1 M:			
			4.5°C	0.14	1,11	5
			35°C	5.7×10^{-3}	1,8	5
Chymotrypsinogen	25 600	SE	pH 7.9, $I = 0.03$:			
			25°C	0.091	1,2	6
			25°C	0.084	2,3	6
Chymotrypsinogen	25 600	SE & GF	pH 3.86, $I = 0.2$	0.4	1,2	7
Chymotrypsinogen	25 600	SE	pH 4.1, 0.1 M NaCl	2.15×10^4	1.2	8
TMV protein	17 000	SE	pH 7.0, $I = 0.1$, 20°C	13.3	1,3	9

Phosphoglycerate kinase	41 000	SE & SV	pH 7.0, 20°C	0.01	1,2	10
Phosphoglycerate mutase	27 500	SE & SV	pH 7.0	40–300	1,4	11
Insulin	5800	SE	pH 2.9, $I = 0.1$:			
			15°C	15.5 ± 0.38	1,2	12
			15°C	2.16 ± 0.12	2,4	12
			15°C	0.036 ± 0.026	2,4,6[c]	12
Ovalbumin +	40 000	GF	pH 6.8, 4°C			
myoglobin	17 816			0.16	—	13
Ovalbumin +	40 000	SE	pH 6.8, 25°C	81	—	14
lysozyme	13 000					
α – chymotrypsin	23 000	SV	pH 3.8	0.091	1,2	
			pH 3.8	0.22	2:1,3[d]	15
Myosin	458 000	QLS	0.5M KCl, 0.2M PO_4, 0.01 M EDTA, pH 7.3	0.13–1.06	1,2	16
Glucose phosphate dehydrogenase	$S_{20,w}$ 0.68 ps Figure 2.2	UCV	pH 7.0, 20°C	1.84×10^3	2,4	17

[a] SE = sedimentation equilibrium; SV = sedimentation velocity; LS = light scattering; QLS = quasi-elastic light scattering; GF = gel filtration; UCV = ultracentrifugation velocity.
[b] Dimensions for dimerization.
[c] $L = w_6/w_2w_4$ (w = concentration in $g\,l^{-1}$).
[d] $L = w_3/w_1w_2$.

References: 1, Visser *et al.* (1972); 2, Timasheff and Townend (1961); 3, Cohen *et al.* (1976); 4, Sophianopoulis and van Holde (1964); 5, Williams *et al.* (1972); 6, Hancock and Williams (1969); 7, Winzor and Scheraga (1964); 8, Aune and Timasheff (1971); 9, Durham and Klug (1972); 10, Spragg *et al.* (1976); 11, Spragg *et al.* (1977); 12, Jeffrey and Coates (1966); 13, Gilbert and Kellett (1971); 14, Howlett and Nichol (1973); 15, Rao and Kegeles (1958); 16, Herbert and Carlson (1971); 17, Fuller Noel and Schumaker (1973).

from this list that, of all the macromolecules that form a biological cell, only the one type, proteins, undergo reversible associative reactions, but evidence exists to suggest that similar self-associations occur between nucleic acids (see, for example, Figures 3.3 and 3.4). Pearce *et al.* (1975) also showed that 16S ribosomal nucleic acid undergoes reversible association. These interactions have been less studied than for proteins. The list not only includes results from multisubunit complexes but also from protein and peptides, which are normally considered to exist as single chains. As the results show, these macromolecules can be induced to self-associate if conditions are correct, albeit that these conditions are often not physiological (e.g. insulin and β-lactoglobulin at pH 2 or tyrocidines in 30% ethanoic acid), but they reflect solubility changes caused either by induced abnormal charges on the macromolecule or its insolubility in the solvents. Thus, one can divide the studies roughly into those on solvent–solute interactions (insulin, β-lactoglobulin, and tyrocidines) and those that relate structure to biological activity (glutamic dehydrogenase, chymotrypsin, phosphoglycerate mutase, TMV protein, and haemoglobin; Table 4.2). Both groups provide

Table 4.2 Reported values for the association coefficient of haemoglobin (taken from Wilicox, 1978)

System[a]	Conditions	Method[b]	$L_{2,4}$ ($l\,g^{-1}$)	References
Human HbO_2	pH 7.0, 0.09 M NaCl, 0.1 M tris, 20°C	SE	22	1
Human HbCO	pH 7.0, 0.2 M NaCl, 20°C	OP	12	2
Human H6CO	pH 7.0, 0.2 M phosphate, 20°C	GF	16	3
Human HbCO	pH 7.0, 0.02 M NaCl, 25°C	SV	25	4
Human HbCO	pH 7.0, 0.1 M phosphate, 20°C	SV & SE	33–62	5
Sheep HbCO	pH 7.2, 0.2 M phosphate, 5.5°C	OP	62	6
Sheep HbCO	pH 6.9, 0.2 M NaCl 5.5°C	OP	31	6
Sheep HbCO	pH 7.2, 0.2 M phosphate, 20°C	SE	19	7
Sheep HbCO	pH 7.2, 0.2 M phosphate, 5°C	SE	78	7
Human HbCO	pH 7.0, 0.1 M Na^+ (0.09 M Cl′ + phosphate), 3°C	GF	20	8
Human HbCO	pH 6.0, 20°C	SE	120	9
Human HbCO	pH 7.0, tris + 0.1 M NaI, 20°C	SE	54	10
Human HbCO	pH 7.0, 0.05 M $(NH_4)_2HPO_4$, 20°C	SE	62.5	11
Human HbO_2	pH 7.0, $I = 0.1$, 0.05 M phosphate:			
	20°C	SE	19.7	12
	6.5°C	SE	46.3	12

[a] HbO_2 = oxyhaemoglobin; HbCO = carboxyhaemoglobin.

[b] The symbols for the methods are the same as in Table 4.1 with the addition of OP = osmotic pressure.

References: 1, Kellett (1971); 2, Guidotti (1967); 3, Ackers and Thompson (1965); 4, Kirshner and Tanford (1964); 5, Edelstein *et al.* (1970); 6, Johnson and Perrella (1971a); 7, Johnson and Perrella (1971b); 8, Chiancone *et al.* (1968); 9, Hensley *et al.* (1973); 10, Williams and Kim (1976); 11, Crepeau *et al.* (1974); 12, Spragg *et al.* (1977).

information on the energies of interaction between subunits, and relate to the view that active macromolecules exist in solution by balancing solute–solvent interactions with biological activity.

Of the techniques available for studying these associations, the ultracentrifuge and light scattering are the most popular. Intensity light scattering can only be applied for studying solutions in which reasonable proportions of free subunits occur in concentrations greater than $1\,\mathrm{g\,l^{-1}}$. Lower concentrations can be studied with the ultracentrifuge, where, in one experiment, the range of concentrations produced in one experiment can extend down to less than $1\,\mathrm{g\,l^{-1}}$ (see Chapter 2). Naturally, results produced from the analyses of a single experiment are more dependent on the accuracy of these data than when several experiments are combined (see Chapter 2 for a discussion of errors). Of the two modes of operating the ultracentrifuge, the equilibrium mode has been more widely applied for studying macromolecular interactions, so this will be considered first.

The basic equations for the equilibrium mode were introduced earlier in this chapter (eq. 4.6), and these are converted to operational equations through two procedures. (*a*) The concentration distribution is converted to a series of point-average relative masses at discrete radii, and from these are calculated the interaction coefficients (as in Adams, 1965, 1967; Roark and Yphantis, 1969; eq. 3.38). (*b*) The individual concentrations of each species in the distribution are calculated from an integral equation (eq. 2.25; as in Sophianopoulis and van Holde, 1964; Reinhardt and Squire, 1965; Haschemeyer and Bowers, 1970; Roark, 1976; Spragg *et al.*, 1977). An example of the first procedure is given by Tang and Adams (1976) where the point averages are related to the interaction and virial coefficients for dimerization by

$$\frac{M_1}{\bar{M}^{\mathrm{app}}_{\mathrm{w}\,(c)}} = \frac{M_1}{\bar{M}_{\mathrm{w}\,(c)}} + B_2 M_2 c = \frac{1}{N - f_1(N-1)} + B_2 M_1 c \qquad 4.13$$

where N is the stoichiometry of the reaction (2 for dimerization), f_1 is the fraction of monomer, and $M^{\mathrm{app}}_{\mathrm{w}\,(c)}$ is the apparent weight-average relative mass estimated from $\mathrm{d}\ln c/\mathrm{d}(r^2)$ at c ($\bar{M}_{\mathrm{w}\,(c)}$ is the true relative mass). Unlike eq. 4.6b, eq. 4.13 provides a linear equation relating $M_1/M^{\mathrm{app}}_{\mathrm{w}\,(c)}$ to c from which $B_2 M_1$ can be calculated from the slope, and either N or f_1, but not both, can be calculated from the intercept ($c = 0$). Although it is theoretically advantageous to include non-ideality in the mathematical model, eq. 4.13 only goes some way to doing that since it assumes that the single virial coefficient applies to both monomer and dimer (as in eq. 4.6). As Reisler and Eisenberg (1971) show, this is not acceptable for large aggregates (reviewed in detail by Eisenberg, 1976). Nevertheless, this model is satisfactory for many systems (for example, β-lactoglobulin; see Table 4.1), but false values for B_2 and f_1 (or N) will occur if estimates of $M^{\mathrm{app}}_{\mathrm{w}\,(c)}$ contain systematic errors. For example, if due allowance is not made for density changes along the column run at high speeds, the result would be a lower value from the intercept (low f_1 or high N) than was the truth (see Chapter 2, Figure 2.3). At low speeds, it is reasonable to assume constant density, but here the ratio of $c(\mathrm{b})/c(\mathrm{m})$ is small.

making the relationship between $\ln c$ and r^2 linear within the errors. Hence, eq. 4.13 cannot be applied to estimate f_1 (or N) if the speed is too low.

In the second method, the concentrations of the individual species are related to the centrifugal variables through equations like eq. 2.25. Unfortunately, applying an expression containing sums of individual Rinde equations (eq. 2.25) is theoretically unsound because this sum can only be formed if the species are non-interacting. This means that if c_i° (the static concentration of the ith species in eq. 2.25) is calculated from results obtained for an interacting system, then the values are not true static concentrations. Several workers have shown that this c_i° can be corrected for interaction by allowing for the rotor speed when calculating the static values (Sophianopoulis and van Holde, 1964; Adams, 1967; Reinhardt and Squire, 1965; Spragg *et al.*, 1977). Interaction coefficients calculated from these corrected concentrations are then indistinguishable from those estimated using point averages for the relative masses (Spragg *et al.*, 1977). These corrections are not necessary if eq. 2.15b is integrated without assuming conservation of mass, to give eq. 2.24. For a mixture of solutes, each solute will distribute according to eq. 2.24; thus, the complete equation will be a sum of the right-hand sides of eq. 2.24. The result of fitting this equation to an experimental record is estimated concentrations ($c(r_0)$, eq. 2.24) for each species at radius r_0. These can then be combined to give an interaction coefficient. This method is exact for ideal solutions, but has the disadvantage that the solutions cannot be checked for loss of solute caused by denaturation, whereas the method in which static concentrations are estimated does provide a route for calculating the initial concentrations. Non-ideality can be introduced through using the second virial coefficient to modify N (eq. 2.24) in the argument of the exponential, but it is unlikely that methods using either eqs. 2.25 or 2.24 could provide an estimate of B_2.

One interesting deviation from the basic trend to analyse equilibrium results through the Rind equation comes from Nichol *et al.* (1976). This procedure does not involve solutions of simultaneous equations involving sums of exponentials nor differentiation of experimental results to give weight-average relative masses. Instead, it works through the definition of dimensionless variables for each species, each of which resembles in form the integrated version of the equilibrium equation (eq. 2.24) divided by the concentration chosen at selected radii: these variables are essentially differences in slope of the measured concentration between two radial positions. The values of these variables are calculated at as many positions as pairs of points can be reasonably identified, and the values of the individual concentrations of the component species are determined by iteration with the experimental data.

All these methods require accurate estimates of baselines in order to correct records, although using only approximate values for M_1 will not materially change the estimates for the interaction coefficients (see Section 4.2.1). Instrumental and environmental errors probably account for the spread of results reported from different laboratories for the same system. For example, with oxyhaemoglobin, the range recorded for apparently similar conditions (0.1–

0.2 M phosphate, 20°C) is from 62 to 16 l g^{-1} (Table 4.2). It is unlikely that doubling the phosphate at this level would produce such a marked decrease in $L_{2,4}$. Equally important for reproducibility is care in maintaining experimental conditions, because different buffer constituents can disturb results. Even if pH is maintained, the type of buffering ions can affect the interaction. This has been illustrated for oxyhaemoglobin in Chapter 1 (Figure 1.4) where substituting ethanoate for phosphate changed the sedimentation characteristics. Whether this also changed the association is not recorded for this particular case, but a complete analysis of these results could only be made if the true buoyant density of the hydrated macromolecule had been determined in the two solutions (see Section 4.3.1, for similar problems when discussing LDH).

Since the lower limit of concentrations usable for light-scattering experiments is high (> 1 g l^{-1}) relative to that of the centrifuge (down to 40 mg l^{-1}), the range of coefficients that can be measured by light scattering with moderately sized macromolecules is limited. However, if the concentration of subunits produced by dissociation of large macromolecules is reasonable, then both intensity and quasi-elastic light scattering can give accurate independent estimates of association coefficients. Light scattering is particularly useful for studying interaction between large aggregates as in the association of myosin (Figure 4.4).

Previous results show that the interaction between macromolecules can be moderated by the type of diffusible ion or ligand present in the solution. These ligands form complexes with the macromolecules which have biological importance in enzyme reactions, and the resulting complex can have interaction coefficients which are different from the uncomplexed molecule. Thus, when discussing association–dissociation reactions, the influence of these small molecules on the reaction must be considered. The energies of the reversible binding of these small molecules often fall in the category of moderate energies, and their influence makes the gross results listed in Tables 4.1 and 4.2 more complex than shown by the estimate of the macromolecular association coefficient. Binding of ligands can be studied through equilibrium dialysis, when the ligand concentration in the dialysate is measured before and after reaching equilibrium. Since the equilibration involves diffusion of the ligand across the membrane, the process takes time to reach equilibrium. A faster procedure is to pass the macromolecule down a gel column previously equilibrated with the ligand and to measure the amount of ligand absorbed. From the results can be calculated the interaction coefficients. Using this method, Fairclough and Fruton (1966) measured the binding of tryptophan and its derivatives to bovine serum albumin. The magnitude of these interaction coefficients put them in the categories of strong and medium interactions ($L = 955$ l g^{-1} for tryptophan and 11 g^{-1} for acetyltryptophan). These examples of absorption of ligands by a protein may not represent the normal for all proteins because bovine serum albumin is noted for its scavenging properties. Nevertheless the range of the binding illustrates the potential of a macromolecule for attracting small molecules. In some macromolecules, the binding of a particular ligand will be higher than others, suggesting that it is fitting into a specially constructed site.

Reversibility of an associative reaction is inherently assumed for all the analyses of interactions. Normally, this is implied in the studies, but experimental proof is more satisfying. This can be obtained experimentally either by diluting or by concentrating a sample after measuring its relative mass and then re-estimating the mass; results from one concentration seldom prove reversibility of a reaction. These lengthy experiments with natural macromolecules often lose their point if traces of hydrolytic enzymes are present. Cassman and Schachman (1971) found that commercial samples of glutamic dehydrogenase were contaminated with a proteolytic enzyme which slowly digested the protein during experiments. In this case, the hydrolytic activity was removed by pretreating the sample with thioethanol and EDTA before studying the dissociation. Durham (1972) also showed that contamination of TMV protein by proteolytic enzymes was the cause of instability in stored samples of coat protein. Observations on dilute samples of bovine serum albumin ($< 1\ \mathrm{g\,l^{-1}}$) showed that these are degraded to form insoluble products within a few days at 4°C. For these reasons, it is important to hasten equilibrium when using the centrifuge by the technique of overspeeding (Section 2.2). For the same reasons, it is worth collecting multiple records covering a range of concentrations run simultaneously in one experiment.

With some systems, it is possible to state the stoichiometry of the reaction with some certainty, but generally this must be determined experimentally. An example where denaturation suggested erroneous stoichiometry was discussed earlier for oxyhaemoglobin (Section 4.3), and other cases exist where it is uncertain what the limiting size of the aggregate may be (β-lactoglobulin, under different conditions to those shown in Table 4.1 (McKenzie, 1971) or insulin or TMV protein). Hence, experiments must be designed to determine the stoichiometry. Even where limiting sizes of the complexes are known, there remains the uncertainty whether intermediaries can be detected during the polymerization. Jeffrey and Coates (1966) report values of coefficients for insulin which include monomer–dimer, dimer–tetramer, and dimer plus tetramer to give hexamer. These multiple reactions contrast with results for proteins like phosphoglycerate mutase where no intermediaries between monomer and tetramer were detected (Spragg *et al.*, 1977). No simple rule can be suggested for measuring these stoichiometries using hydrodynamic methods. Normally, the results are based on statistical tests for the best model to fit the record. Usually, methods which employ equilibrium are not sensitive tests for deciding on the stoichiometry, but often transport experiments give better identification. Naturally, the method discussed on p. 166 using cross-linking reactions is the most definitive, provided that the conditions for the cross-linking match those studied in the hydrodynamic method (see Thomas and Butler, 1977, for an example of this test).

Analysing transport experiments through average parameters (Section 4.2.1) makes it a relatively insensitive method for cases where low concentrations of small subunits occur. Thus, these results can only confirm or cast doubt on equilibrium results, but they do little to aid the determination of stoichiometry. A more useful method for analysing transport experiments is to analyse the shape of the transported boundary. Detailed instructions of the procedure have been given by Gilbert and Gilbert (1973) (these have been discussed in Section 4.2.2). For the analysis to be successful, within the model it is necessary to include hydrodynamic parameters such as concentration dependence (g, eq. 2.22) or the equivalent for electrophoresis (Cann, 1970) and gel chromatography (Chiancone *et al.*, 1968). Ignoring the concentration dependences of hydrodynamic parameters can produce false impressions of the dependence of $S_{20,\mathrm{w}}$ on concentration (as with older analyses of oxyhaemoglobin; Wyman, 1963). If the values of the interaction coefficients differ significantly between the experimental procedures (as with yeast phosphoglycerate mutase; Spragg *et al.*, 1977), then the sources of the anomaly must be traced. These may reflect slow rates of reaction or false stoichiometries or traces of additional small components which may or may not be important to the association. In sedimentation equilibrium, small subunits would not be included in the model, but would raise the apparent concentration of the monomer in the reaction to give a lowered estimate of the interaction coefficient.

4.3.3 Weak interactions

The division between selective weak associations of biological significance and solvent–solute non-ideal interactions is often uncertain, and, even if differences can be defined theoretically, it is difficult to separate the two experimentally. Measurable effects caused by weak interactions between macromolecules can only be produced by working with high concentrations of macromolecules. If $L_{1,2} < 2 \times 10^{-2}\,\mathrm{l\,g^{-1}}$ to give a $\Delta G/RT < 4$, then it would be a debatable whether it would be possible to separate selective association from non-ideality (Section 4.2.1). An $L_{1,2}$ of this magnitude would produce 50% association at $100\,\mathrm{g\,l^{-1}}$, and the association would be compensated for by lowered relative mass caused by non-ideality. In other words, there would be no observable concentration dependence of the estimates of mass; this could be confused with working at a theta condition. For even smaller levels of interaction, concentration dependence would still be observed, but the calculated B_2 would be less than expected for a spherical molecule. The usable levels may vary between transport and equilibrium experiments, but the above comments apply to all methods of analysis. It is for these reasons that theoretical attempts to calculate concentration dependence from molecular coordinates are particularly important.

The fact that it is difficult to measure weak interactions does not mean they are not important biologically. For example, if weak association could order a sequence of enzymes to give a loose complex, then the reaction rates would be different from those measured in free solution. Studying interactions of this type is made difficult because of the many variables introduced by having more than one molecular type in solution. Even when coefficients for the interactions belong to the medium classification, there still remains the introduction of more than one unknown concentration dependence parameter (with dimerization, simple assumptions reduce this to one; see Section 4.1). Two examples of this mixed reaction are given in Table 4.1 for the reaction between ovalbumin and myoglobin or lysozyme; these energies belong to the medium interactions, but it is unknown whether these particular reactions have any biological significance. A further expression of long-range energetically weak association occurs during the selection of antigens by preformed antibodies. The rate of selection may be slow in term of rates of chemical reactions but, once recognition has occurred and the two macromolecules are within the range of the attractive forces, energetically strong links form rapidly to produce a strongly linked complex. The selection resembles, in many respects, the nucleation of crystals from a mixture, where initially several molecules must come together to form an energetically favoured complex. This process must involve balancing repulsion by thermal agitation (for uncharged molecules) against selective net attraction between like molecules. The lower limit for forming a stable complex occurs when the energies of interaction are greater than thermal energy (Brownian motion). This energy moves the molecules apart and prevents the formation of a stable aggregate. This means that dimerization between macromolecules having molar interaction coefficients of less than 0.37 $\mathrm{M\,l^{-1}}$ (giving an $L_{1,2}$ for a macromolecule of relative mass 60 000 daltons of $1.2 \times 10^{-5}\,\mathrm{l\,g^{-1}}$) is forbidden at normal temperatures. The difference between this lower limit and the existing measurements shows that there is plenty of scope for these interactions to go undetected.

The experimental necessity to compete with hydrodynamic and thermodynamic non-ideality imposes severe limitations on classical techniques for studying weak interactions. Thus, with the sole exception of gel chromatography, none of the equilibrium or transport procedures discussed in Chapters 2 and 3 is suitable for measuring directly weak associations. Gel chromatography is possibly the most powerful procedure available at present for these studies when applied in the differential mode (as in Figure 2.7). It may be possible to extend this mode so that absolute coeficients can be measured; this requires a suitable rearrangement of the experimental procedure. Certainly it is possible with this method to measure differences in interactive energy as low as 1.11 $\mathrm{kJ\,mol^{-1}}$ (Gilbert and Gilbert, 1973, and shown in Figure 2.7). Working at these limits requires careful attention to column construction, measurement of elution volumes (determined by weighing) and accurate determination of optical absorbances of the eluate. These factors have deterred all but the enthusiast from this field of low energy interactions.

A less direct approach to the problem may be possible through the interpretation of the concentration dependence of hydrodynamic parameters (that is, diffusion coefficients and sedimentation coefficients, see Chapters 2 and 3). This is particularly important when discussing encounter distances in kinetics. The level of the dependence reflects not only the excluded volume of the macromolecule but also the frictional concentration dependence introduced through solute–solvent–solute interactions. The energy of these interactions can be estimated if suitable physical models existed to make it possible to transform the experimental measurements. A basis for these transformations exists in the Anderson and Reed (1976) model for diffusion (Section 3.3.2) and the Goldstein and Zimm (1971) model for sedimentation (Section 2.1.3).

4.4 Dynamic relaxation

The concept of equilibrium as a probability function defining the concentration of components in one state compared with a second leads naturally to the view of dynamic interchange between the states. This dynamic approach can be used for studying macromolecular interactions by perturbing the equilibrium condition and monitoring the temporal relaxation back to the equilibrium. In order to produce this relaxation, the value of one or more properties of the subunits must be different when it is in the complex compared with when it is free. Possibly, the two properties most widely employed are changes in polarizability as reflected by partial molar volumes and changes in total free energy caused by thermal perturbations. In the first case, changing the external pressure will cause readjustment of the equilibrium, and this is achieved experimentally by bursting a diaphragm separating the high hydraulic pressure from the solution (Davies and Gutfreund, 1976; Schulz *et al*., 1976). In the second case, a rapid temperature jump is achieved by discharging a capacitor through the solution (Eigen, 1968). Both methods require that the perturbations should be completed in a timescale which is considerably less than that of the reaction rates producing the equilibrated condition; some overlap can be deconvoluted mathematically. Fortunately, this problem does not limit the application to macromolecules, because there is about 10^{-4} s available for bursting the diaphragm or discharging the condenser, and diffusion over the reaction coordinates requires longer periods. A contrasting method, but one which yields similar results, is the so-called stop-flow method where the reactants are mixed rapidly in a chamber and the progress of association monitored.

Having disturbed the equilibrium, many optical procedures have been devised to monitor the formation of the complex (see Eigen, 1968, for a review), but possibly the most common employ spectrophotometric methods (which depend on the coefficient of absorbance changing during the reaction), fluorescence changes, and monitoring the turbidimetric or light-scattering changes. All methods produce a decay (or rise) in the property, and this is recorded on an oscilloscope, giving records similar to those shown in Figure 4.7. From these curves can be estimated the relaxation time of the reaction (τ, corresponding to the definition of relaxation discussed in Section 3.2.2) which numerically equals the time for the change to drop (or rise) to $1/e$ of the value at the beginning of the experiment. Eigen (1968) formulates the relationship and reaction rates for many types of reactions, but, continuing the example of dimerization (eq. 4.1), a simple relationship can be derived. The rate equations for eq. 4.1 are

$$d[A]/dt = -k_1[A]^2 + k_2[A_2] \tag{4.14}$$

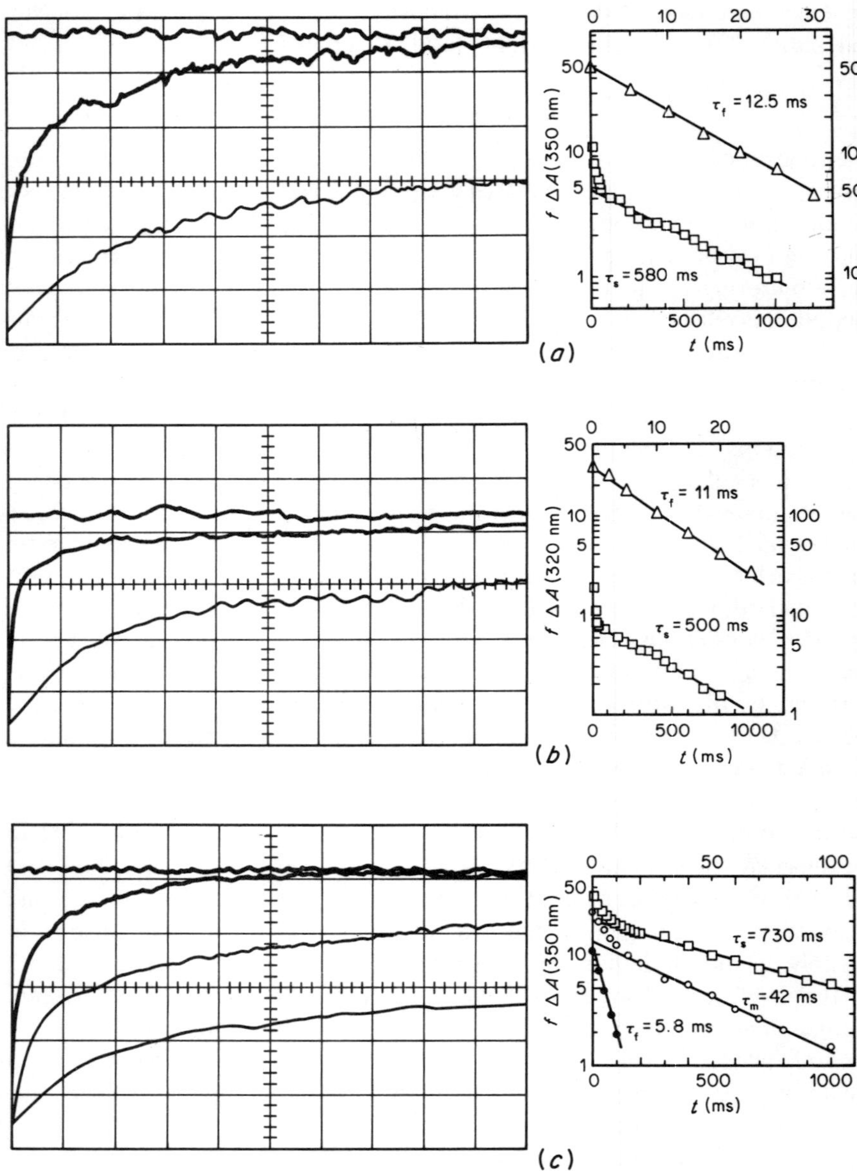

Figure 4.7 Oscilloscope recordings taken from a pressure-jump relaxation experiment made with *E. Coli* ribosomes: (*a*) 30S subunits, (*b*) 50S subunits, and (*c*) a mixture of 30S and 50S subunits. (Taken from Shulz *et al.*, 1976, *Biophys. Chem.* **11** 253, Figure 1. Reproduced by permission of North-Holland Publishing Company)

which when linearized becomes

$$dx_1/dt = \xi_1 x_1 + \xi_2 x_2 \tag{4.15}$$

in which

$$x_1 = \delta[\mathrm{A}] \qquad x_2 = \delta[\mathrm{A}_2] \tag{4.16a}$$

$$\xi_1 = -k_1[\mathrm{A}]^2 \qquad \xi_2 = k_2 \tag{4.16b}$$

Thus, after the disturbance and before equilibrium has had time to relax from the original state, $dx/dt = 0$. Hence, during relaxation, $(\xi_1 + \tau)$ and ξ_2 are the eigenvalues for this linear equation (eq. 4.15) and must be estimated from the decay. In this example, the rates can be calculated by substituting for $1/t$ in the conventional rate equations to give

$$\frac{k_1[\mathrm{A}]}{y} = \tau \ln\left(\frac{y}{y-z}\right) \tag{4.17a}$$

$$k_1 + k_2 = \tau \ln\left(\frac{y}{y-z}\right) \tag{4.17b}$$

where y is the maximum change in concentration of A at the new equilibrium (should be at infinite time, but normally a value can be taken after an acceptably long time), while z is the concentration of A at time t. When intermediaries are formed in the reaction (as in the self-association of insulin, Table 4.1), then several relaxation times will be recorded, and the result is a sum of exponentials in which τ_i appears in the arguments. Resolving these multiple exponentials is not a simple task and, in practice, this limits the application of the method (see de Maeyer *et al.* (1976) for a full discussion of both the thermodynamics of the fluctuations and their treatment).

Pecht and Lancet (1977) have reviewed the relationships expected for macromolecular interactions of the type found in antibody–antigen reactions, and generally described by

$$\mathrm{A} + \mathrm{B} \underset{k_{-1}}{\overset{k_1}{\rightleftharpoons}} \mathrm{AB} \underset{k_{-2}}{\overset{k_2}{\rightleftharpoons}} \mathrm{C} \tag{4.18}$$

The complex AB is formed through an encounter step, while the second step is the actual binding step. Since both steps are involved in associative reactions, but only the encounter step is the diffusion-controlled part, it is possible to relate k_1 and k_{-1} to the diffusion coefficients (D_i) of the macromolecules. Eigen (1974) has shown that for neutral macromolecules the constants of the encounter step are given by

$$k_1 = 4\pi\mathcal{N}(D_\mathrm{A} + D_\mathrm{B})r_\mathrm{AB} \tag{4.19a}$$

$$k_{-1} = 3(D_\mathrm{A} + D_\mathrm{B})/r_\mathrm{AB}^2 \tag{4.19b}$$

In eq. 4.19, r_AB is called the encounter distance, and normally is difficult absolutely to define experimentally. It does seem, however, that this distance determines in part the experimental observations that there are concentration dependences for diffusion and sedimentation coefficients (Section 3.3.2), and r_{AB} could in principle be estimated from these observations using models such as those given by Goldstein and Zimm (1971), Phillies (1975, and eq. 3.13), and Anderson and Reed (1976). These models include a term from statistical mechanics for the relationship between energy of interaction ($w(r)$ in eq. 3.13) between like molecules and the distance between the molecules—the form of the relationship is

$$\int A(r)\exp[-w(r)/kT]\mathrm{d}r$$

where $A(r)$ is a hydrodynamic coefficient. Thus, a repulsive peak in the total-energy profile (this is the algebraic sum of attraction and repulsion) would indicate the distance out from the surface of the molecule at which an energetic barrier occurs (usually, the barrier is less than unity when normalized to kT, but even small barriers would select molecules of greater energy than the average). This approximate distance when added to the radii of the molecules would provide an approximate value for r_{AB}. Using this link between eq. 4.19 and concentration dependence of diffusion coefficients is educationally useful since it indicates the molecular factors that modify the kinetics of the encounter step. For example, a change in the balance between the charge on the molecule (or a redistribution of the charge around the molecule) and/or a change in the molecular polarizability (which determines attraction) would only require a change of one or two amino acids in the chain. To date, numerical difficulties have made it impossible to pursue this approach further than qualitative concepts, but these concepts have a basis in experimental observations of the concentration dependence of diffusion and sedimentation coefficients.

The true values for the complete encounter step will depend on whether all the sites on the spherical molecule are equally probable for reaction (and calculations of concentration dependence assumes there is no preferred direction). If the binding can only occur at limited points, then the values calculated by eq. 4.19a must be reduced by an orientation factor usually assumed to be of the form $2\pi[1-\cos(\alpha/2)]$ where α is the apex angle of a solid cone. Thus, when $\alpha = 180°$ (a flat plane) the factor becomes 2π, while for $\alpha = 45°$ only 3.8% of the sphere is suitable for binding.

Apart from statistical limitations, there are other fundamental limitations on both the pressure and temperature jump procedures. The origin of their effects can be seen by examining the thermodynamic interrelationships between the two conditions as defined by

$$\mathrm{d}E = \left(\frac{\partial E}{\partial V}\right)_T \mathrm{d}V + \left(\frac{\partial E}{\partial T}\right)_V \mathrm{d}T \qquad 4.20$$

where E is the energy of the system, V is the volume, and T is the temperature. Since work is done on the solvent when relaxing the pressure or discharging the capacitor ($\mathrm{d}V$), this is expressed as a change in temperature ($\mathrm{d}T$). Thus, in theory, it is impossible to separate the molecular source of the perturbations. Fortunately, the equilibrium can be perturbed by relatively small changes in pressure and temperature; for example, the changes of pressure used are about 150 bar (Davies and Gutfreund, 1976) while the temperature change is seldom more than 1°C. The pressure change may seem large enough to raise the temperature by 1°C, especially when it is remembered that water (the usual solvent) is relatively incompressible. However, this effect is offset by the high specific heat of water, which thus absorbs the energy and prevents a large rise in temperature. Order-of-magnitude calculations show that an adiabatic pressure change of about 150 bar would raise the temperature by less than 0.1°C. Conversely, this value shows that a 1°C change in temperature only raises the pressure by about 15 bar.

In general, these relaxation techniques have been more widely applied in kinetics of enzyme reactions than in studying association–dissociation reactions. This is because macromolecular dissociations are not very sensitive to temperature changes (entropy effects tending to balance rises in temperature; see hydrophobic effects, Chapter 1), while changes in partial specific volume accompanying association are usually small. Davies and Gutfreund (1976) quote a value of 26 $cm^3\,mol^{-1}$ for the $\Delta\bar{V}$ of the self-association of tubulin, and surprisingly this value is similar in magnitude to the $\Delta\bar{V}$ of the second dissociation of phosphate ions. Hence, even if small recorded changes could be observed in association of macromolecules in phosphate buffer, interpreting them requires that allowance be made for pH changes. Larger values for $\Delta\bar{V}$ are recorded for the self-assembly of myosin to form filaments (– 280 $cm^3\,mol^{-1}$) and these changes produce anomalous sedimentation results with myosin (Josephs and Harrington, 1968) where pressures of about 250 bar can be generated in the ultracentrifugal cell. Similar effects appear to occur during the flotation of self-associating low-density lipoproteins (Figure 2.4), but this system has not been studied extensively. When subunits associate to produce large complexes, $\Delta\bar{V}$ appears to increase in magnitude, so making it possible for Schulz *et al.* (1976) to study the association of 30S and 50S ribosomes using the pressure jump technique (Figure 4.7). From analyses of these curves, they were able to define three relaxation times for the reactions, 5–10 ms, 40–100 ms, and 500–1000 ms. From a study of the influence that Mg^{2+} had on the reaction, they concluded that the reaction

$$30\text{S} + 50\text{S} \rightleftharpoons 70\text{S}$$

had a relaxation time of between 40 and 100 ms. The other two relaxation times were thought to be caused by self-association of the form

$$n(30\text{S}) \rightleftharpoons (30\text{S})_n$$

and

$$n(50\text{S}) \rightleftharpoons (50\text{S})_n$$

but they were unable to separate these multiple reactions in order to estimate association coefficients.

The stop-flow method has largely been unexplored as a technique for studying the kinetics of association of macromolecules, largely because of suitable techniques to monitor the changes occurring during the reaction. The only general change that occurs is an increase in turbidity as the polymers form, so it is interesting to find that laser light scattering (Chapter 3) has been enlisted for monitoring this type of reaction (Liddle *et al.*, 1977). In order accurately to monitor the changes in relative mass on a timescale of about 20 ms, means that the optics and electronics must be carefully aligned to maintain an optimum signal-to-noise ratio. Results show that $\Delta\bar{M}_w$ as low as 2×10^5 dalton can be measured using this equipment (Liddle *et al.*, 1967), while the decay in the equilibrium state can be measured with acceptable precision (the dead time of the equipment is about 1 ms, making it suitable for studying self-association of most macromolecules).

4.5 Conclusions

The results of studies made on 'sticky molecules' show that association–dissociation reactions are common properties for macromolecules in solution. The range of energies encountered is wide, but unfortunately our knowledge of the importance that these have in maintaining stability of biological systems is set by the limits of experimental techniques. Hydrodynamic techniques are largely confined to studying medium-energy interactions, although differential methods mean that it is possible to detect and measure small differences in the association–dissociation of these reactions. With these methods, the problem is in interpreting results, since a wide knowledge of the behaviour of macromolecules in solution is required. Possibly, developing techniques using bifunctional reagents combined with gel methods may extend and simplify the qualitative measurements. It is in the region of weak interactions that confusion caused by non-ideal states of solutions makes measurements particularly difficult, but whether 'freezing' the equilibrium with the reagents becomes possible (and so reduces this problem) remains to be seen.

Even if new techniques make it possible to cover wide-ranging energies of interaction, these data will only be useful if they serve to test underlying principles on macromolecular interaction. It is difficult to see what the principles are that force some species of organism to produce multisubunit enzymes when the same catalytic action can be achieved by simpler complexes in other species. To say that these differences between species (or organs, in some cases) reflect solvent conditions only rephrases the question, but does not provide a universal hypothesis. Cooperativity is one guiding principle, but again this activity appears to apply to only a limited few oligomeric complexes. Possibly, there is not a universal principle. However, the inconsistency in size found between macromolecules having the same enzymic activity but differing origin leads one to believe that often the complexity is not required for the known biological activity but is the outcome of some other biological requirement.

References

Ackers, G. K. (1970) *Adv. Prot. Chem.* **24**, 343.

Ackers, G. K. (1975). In *The Proteins*, Vol. 1, Eds. H. Neurath and R. L. Hill, Academic Press, New York, p. 2.

Ackers, G. K. and Thompson, T. E. (1965) *Proc. Natl. Acad. Sci.* (*Washington*) **53**, 342.

Adair, G. S. (1925) *Proc. R. Soc.* A **108**, 627.

Adair, G. S. (1928) *Proc. R. Soc.* A **120**, 573.

Adams, E. T. (1965) *Biochemistry* **4**, 1646.

Adams, E. T. (1967) *Annals N.Y. Acad. Sci.* **164**, 226.

Adelman, S. A. and Deutch, J. M. (1975) *Macromolecules* **8**, 58.

Ahluwalia, J. C., Ostiguy, C., Perron, G. and Desnoyers, J. E. (1977) *Can. J. Chem.* **55**, 3364.

Aksiyote-Benbast, J. and Bloomfield, V. A. (1975) *J. Molec. Biol.* **95**, 335.

Alexander, A. E. and Johnson, P. (1949) *Colloid Science*, Vol. 1, Oxford University Press, Oxford, Chap. X.

Alpert, S. S. and Banks, G. (1976) *Biophys. Chem.* **4**, 287.

Amos, L. A. (1974) *J. Microsc.* **100**, 143.

Anderson, N. G. (1976a) *J. Theor. Biol.* **60**, 401.

Anderson, N. G. (1976b) *J. Theor. Biol.* **60**, 413.

Anderson, J. L. and Reed, C. C. (1976) *J. Chem. Phys.* **64**, 3240.

Andrews, P. (1964) *Biochem. J.* **91**, 222.

Andrews, P. (1965) *Biochem. J.* **96**, 595.

Archibald, W. J. (1947) *J. Phys. & Colloid Chem.* **51**, 1204.

Aune, K. C. and Timasheff, S. N. (1971) *Biochemistry* **10**, 1609.

Baldwin, R. L. (1957a) *Biochem. J.* **65**, 503.

Baldwin, R. L. (1957b) *Biochem. J.* **65**, 490.

Batchelor, G. K. (1972) *J. Fluid Mech.* **52**, 245.

Batchelor, G. K. (1976) *J. Fluid Mech.* **74**, 1.

Berkeley, Earl of, and Hartley, E. G. J. (1906) *Proc. R. Soc.* A **77**, 156.

Bernal, J. D. and Fankuchen, I. (1941) *J. Gen. Physiol.* **25**, 111.

Berne, B. J. and Pecora, R. (1976) *Dynamic Light Scattering*, John Wiley & Sons, New York.

Bier, M. (Ed.) (1959) *Electrophoresis: Theory, Methods and Applications*, Academic Press, New York.

Birnbaumer, M. E., Schrader, W. T. and O'Malley, B. W. (1979) *Biochem. J.* **181**, 201

Bloomfield, V. A. (1966) *Biochemistry* **5**, 684.

Bloomfield, V. A., Dalton, W. O. and van Holde, K. E. (1967a) *Biopolymers* **5**, 135.

Bloomfield, V. A. and De La Torre, J. G. (1977) *Biopolymers* **16**, 1747.

Bloomfield, V. A., van Holde, K. E. and Dalton, W. O. (1967b) *Biopolymers* **5**, 149.

Blow, D. (1974) *Bayer Symposium 5: Proteinase Inhibitors*, Eds. H. Fritz, H. Tschesche L. J. Greene and E. Truscht, Springer-Verlag, Berlin, p. 677.

Brandts, J. F., Olivera, J. and Westart, C. (1970) *Biochemistry* **9**, 1038.

Brenner, S. L. and McQuarrie, D. A. (1973) *Biophys. J.* **13**, 301.
Broersma, S. (1960) *J. Phys. Chem.* **32**, 1632.
Brown, J. R. (1975) *Fed. Proc. Fed. Amer. Soc. Exp. Biol.* **34**, 591.
Bucci, E. (1971) *J. Molec. Biol.* **55**, 401.
Burgers, J. M. (1938) *Second Report on Viscosity and Plasticity, Amsterdam Academy of Science*, Nordemann Publishing Co., Amsterdam, Chap. 3.
Burgers, J. M. (1942) *Proc. Acad. Sci., Amsterdam* **45**, 126.
Camerini-Otero, R. D., Pusey, P. N., Koppel, D. E., Schaefer, D. W. and Franklin, R. M. (1974) *Biochemistry* **13**, 960.
Cann, J. R. (1970) *Interacting Molecules*, Academic Press, New York and London.
Cassassa, E. F. and Eisenberg, H. (1964) *Adv. Prot. Chem.* **19**, 287.
Cassman, M. and Schachman, H. K. (1971) *Biochemistry* **10**, 1015.
Charlwood, P. A. and Utsumi, S. (1969) *Biochem. J.* **112**, 357.
Champagne, M. and Sadron, C. (1954) *Conf. High Polymers, Strasbourg*, as reported in Harrington *et al.* (1956).
Cheng, P. and Schachman, H. K. (1955) *J. Polym. Sci.* **16**, 19.
Cherry, R. J., Burkl, A., Busslinger, M., Schneider, G. and Parish, G. R. (1976) *Nature* **263**, 389.
Chiancone, E. and Gilbert, G. A. (1965) *J. Biol. Chem.* **240**, 3866.
Chiancone, E., Gilbert, L. M., Gilbert, G. A. and Kellett, G. L. (1968) *J. Biol. Chem.* **243**, 1212.
Cho, I. C. and Swaisgood, H. (1973) *Biochemistry* **12**, 1572.
Chotia, C. (1974) *Nature* **248**, 338.
Chu, B. (1967) *Molecular Forces*, John Wiley & Sons, New York.
Chu, B. (1974) *Laser Light Scattering*, Academic Press, New York and London.
Cohen, R. and Hahn, C. (1965) *Compt. Rend.* **260**, 2077.
Cohen, R. J. and Benedek, G. B. (1976) *J. Molec. Biol.* **108**, 151.
Cohen, R. J., Jedziniak, J. A. and Benedek, G. B. (1976) *J. Molec. Biol.* **108**, 174.
Cohn, E. J. and Edsall, J. T. (1943) *Proteins, Amino Acids and Peptides*, Monograph No. 90, American Chemical Society, New York.
Cox, D. J. (1971) *Arch. Biochem. Biophys.* **146**, 181.
Cox, D. J. and Schumaker, V. N. (1961) *J. Am. Chem. Soc.* **83**, 2433.
Craig, M. E., Crothers, D. M. and Doty, P. (1971) *J. Molec. Biol.* **62**, 383.
Crank, J. (1956) *The Mathematics of Diffusion*, Oxford University Press, Oxford.
Creeth, J. M. and Knight, C. G. (1968) *Special Publication No. 23*, The Chemical Society, London, p. 303.
Creeth, J. M. and Pain, R. H. (1967) *Prog. Biophys. & Molec. Biol.* **17**, 219.
Creighton, T. E. (1978) *Prog. Biophys. & Molec. Biol.* **33**, 231.
Crepeau, R. H., Hinsley, C. P. and Edelstein, S. J. (1974) *Biochemistry* **13**, 4860.
Crowther, P. A. and Klug, A. (1975) *Ann. Rev. Biochem.* **44**, 161.
Cummins, H. Z., Knuble, N. and Yeh, Y. (1964) *Phys. Rev. Lett.* **12**, 150.
Cummins, H. Z. and Pike, E. R. (Eds.) (1974) *Photon Correlation and Light Beating Spectroscopy*, Plenum Press, New York.
Danson, M. J., Hale, G., Johnson, P., Perham, R. N., Smith, J. and Spragg, S. P. (1979) *J. Molec. Biol.* **129**, 603.
Darnall, D. W. and Klotz, I. M. (1975) *Arch. Biochem. Biophys.* **154**, 696.
Davenport, W. B. and Root, W. L. (1958) *An Introduction to the Theory of Random Signals and Noise*, McGraw-Hill, New York.
Davies, G. E. and Stark, G. R. (1970) *Proc. Natl Acad. Sci.* (*USA*) **66**, 651.
Davies, J. S. and Gutfreund, H. (1976) *F.E.B.S. Lett.* **72**, 199.
Debye, P. (1929) *Polar Molecules*, Reinhold Publ. Corp. (reprinted by Dover Publ., Inc., New York).
Debye, P. and Hückel, E. (1923) *Phys. Z.* **24**, 185 and 305.
de Maeyer, L., Gnadig, K., Hendrix, J. and Salek, B. (1976) *Q. Rev. Biophys.* **9**, 83.

Dickerson, R. E. and Geis, I. (1969) *The Structure and Actions of Proteins*, Harper & Row, New York.

Dintzis, H. M., Oncley, J. L. and Fuoss, R. M. (1954) *Proc. Natl Acad. Sci.* (*USA*) **40**, 62.

Dishon, M., Weiss, G. H. and Yphantis, D. A. (1966) *Biopolymers* **1**, 449.

Donnan, F. G. and Guggenheim, E. A. (1932) *Z. Phys. Chem.* **162**, 346.

Doty, P. and Bunce, B. (1952) *J. Am. Chem. Soc.* **74**, 5029.

Dunker, A. K. and Kenyon, A. J. (1976) *Biochem. J.* **153**, 191.

Dunker, A. K. and Rueckert, R. R. (1969) *J. Biol. Chem.* **244**, 5074.

Durham, A. C. H. (1972) *F.E.B.S. Lett.* **25**, 147.

Durham, A. C. H. and Klug, A. (1972) *J. Molec. Biol.* **67**, 315.

Dzyaloshinskii, I. E., Lifshitz, E. M. and Pitaevshii, L. P. (1961) *Adv. Phys.* **10**, 165.

Edelstein, S. J. and Gibson, Q. H. (1971) In *Probes of Structure and Function of Macromolecules and Membranes*, Vol. 2, Eds. B. Chance, T. Yonitani and A. S. Mildran, Academic Press, London and New York, p. 417.

Edelstein, S. J., Rehmar, M. J., Olsen, M. J. and Gibson, Q. (1970) *J. Biol. Chem.* **245**, 4372.

Edsall, S. T. and Wyman, J. (1958) *Biophysical Chemistry*, Vol. 1, Academic Press, New York and London.

Eigen, M. (1968) *Q. Rev. Biophys.* **1**, 3.

Eigen, M. (1974) In *Quantum Statistical Mechanics in Natural Sciences*, Eds. S. L. Muntz and S. M. Wiedermayer, Plenum Press, New York, p. 37.

Eisenberg, H. (1971) *Procedures in Nucleic Acid Research*, Vol. 2, Eds. G. L. Canton and D. R. Davies, Harper & Row, New York, p. 137.

Eisenberg, H. (1976) *Biological Macromolecules and Polyelectrolytes in Solution*, Oxford University Press, Oxford.

Eisenberg, H. and Felsenfeld, G. (1967) *J. Molec. Biol.* **30**, 17.

Elliot, A. (1976) *Biochem. J.* **159**, 743.

Evans, J. M. (1972) In *Light Scattering from Polymer Solutions*, Ed. H. B. Huglin, Academic Press, London.

Fair, B. D., Chao, D. Y. and Jamieson, A. M. (1978) *J. Coll. & Interface Sci.* **66**, 323.

Fairclough, G. F. and Fruton, J. S. (1966) *Biochemistry* **5**, 673.

Fermi, G. (1975) *J. Molec. Biol.* **97**, 237.

Fish, W. W., Mann, K. G. and Tanford C. (1969) as quoted in Ackers (1970).

Florey, P. J. (1953) *Principles of Polymer Chemistry*, Cornell University Press, Ithaca, New York.

Ford, N. C. (1972) *Chem. Scr.* **2**, 193.

Fowler, R. H. and Guggenheim, E. A. (1949) *Statistical Thermodynamics*, Cambridge University Press, Cambridge.

Frank, H. S. P. and Evans, M. J. (1945) *J. Chem. Phys.* **13**, 507.

Frieden, C. (1963) *J. Biol. Chem.* **238**, 3286.

Fujita, H. (1961) *Mathematics of Sedimentation*, Academic Press, New York and London.

Fujita, H. and MacCosham, V. J. (1959) *J. Chem. Phys.* **30**, 291.

Fuller Noel, J. K. and Schumaker, V. N. (1973) *Anal. Biochem.* **41**, 1.

Furgusen, K. A. (1964) *Metab. Chem. Exp.* **13**, 985.

Gabler, R. (1978) *Electrical Interactions in Molecular Biophysics*, Academic Press, New York and London.

Gibbons, R. A. (1971) *Biopolymers* **10**, 411.

Gilbert, G. A. (1958) *Die Stärke* **10**, 95.

Gilbert, G. A. and Gilbert, L. M. (1973) *Meth. Enzymol.* **27**, 273.

Gilbert, G. A., Gilbert, L. M. and Shawkey, N. A. F. (1975) *Biochem. Biophys. Acta.* **379**, 488.

Gilbert, G. A. and Kellett, G. L. (1971) *J. Biol. Chem.* **246**, 6079.

Gilbert, L. M., Gilbert, G. A., Owens, C. E. and Shawkey, N. A. F. (1973) *Nature New Biology* **235**, 110.

Godfrey, J. E. and Harrington, W. F. (1970a) *Biochemistry* **9**, 886.

Godrey, J. E. and Harrington, W. F. (1970b) *Biochemistry* **9**, 894.
Goldberg, R. J. (1953) *J. Phys. Chem.* **57**, 194.
Goldstein, B. and Zimm, B. H. (1971) *J. Chem. Phys.* **54**, 4408.
Gosting, L. J. (1956) *Adv. Prot. Chem.* **1**, 429.
Gosting, L. J. and Morris, M. S. (1949) *J. Am. Chem. Soc.* **71**, 1998.
Grant, E. H., Sheppard, R. J. and South, G. P. (1978) *Dielectric Behaviour of Biological Molecules in Solution*, Oxford University Press, Oxford.
Grant, E. H., South, G. P. Takashima, S. and Ichimura (1971) *Biochem. J.* **122**, 691.
Griffith, J. P. (1972) *Biochem. J.* **126**, 553.
Guggenheim, E. F. (1930) *J. Phys. Chem.* **34**, 1751.
Guidiotti, G. (1967) *J. Biol. Chem.* **242**, 3685.
Gurney, R. W. (1953) *Ionic Processes in Solution*, McGraw-Hill, New York (reprinted by Dover Publ. Inc., New York 1962).
Gutfreund, H. (1948a) *Biochem. J.* **42**, 156.
Gutfreund, H. (1948b) *Biochem. J.* **42**, 544.
Gutfreund, H. (1954) *Trans. Faraday Soc.* **50**, 628.
Hamaker, H. C. (1937)*Physica's Grav.* **IV**, 1058.
Hancock, D. K. and Williams, J. W. (1969) *Biochemistry* **8**, 2598.
Harrington, W. F., Johnson, P. and Ottewill, R. H. (1956) *Biochem. J.* **62**, 562.
Haschemeyer, R. H. and Bowers, W. F. (1970) *Biochemistry* **9**, 435.
Heller, W. (1966) *J. Polym. Sci.* A-2 **4**, 209.
Hensley, C. P., Edelstein, S. J. and Gibson, Q. (1973) *Fed. Proc.* **32** (No. 3, Part 1) Abstract 1580.
Herbert, T. J. and Carlson, F. D. (1971) *Biopolymers* **10**, 2231.
Herman, R. B. (1972) *J. Phys. Chem.* **76**, 2754.
Hexner, P. E., Radford, L. E. and Beams, S. W. (1961) *Proc. Natl Acad. Sci.* (USA) **47**, 1848.
Hildebrand, J. H. and Scott, R. L. (1950) *The Solubility of Non-Electrolytes*, 3rd edn., Reinhold Publ. Corp., New York.
Hirschfelder, J. O., Curtis, C. F. and Bird, R. B. (1967) *Molecular Theory of Gases and Liquids*, John Wiley & Sons, New York, Chap. 3.
Howlett, G. J. and Nichol, L. W. (1972) *J. Phys. Chem.* **76**, 2740.
Howlett, G. J. and Nichol, N. W. (1973) *J. Biol. Chem.* **248**, 619.
Hummel, J. P. and Dryer, W. J. (1962) *Biochim. Biophys. Acta* **63**, 530.
Inoue, H. and Timasheff, S. N. (1968) *J. Am. Chem. Soc.* **90**, 1890.
Ip, S. H. C., Johnson, M. L. and Ackers, G. K. (1976) *Biochemitry* **15**, 654.
Ishihara, A. and Hayashida, T. (1951a) *J. Phys. Soc. Japan* **6**, 40.
Ishihara, A. and Hayashida, T. (1951b) *J. Phys. Soc. Japan* **6**, 46.
Israelachvili, J. N. (1974) *Q. Rev. Biophys.* **6**, 341.
Israelachvili, J. N. and Tabor, D. (1972) *Proc. R. Soc.* A **331**, 19.
Jaenicke, R. and Knof, S. (1968) *Europ. J. Biochem.* **4**, 157.
Jamieson, A. M. and Maret, A. R. (1973) *Chem. Soc. Rev.* **2**, 325.
Jeffrey, P. D. and Coates, J. H. (1966) *Biochemistry* **5**, 489.
Jennings, B. R. (1973) In *Light Scattering from Polymer Solutions*, Ed. H. B. Huglin, Academic Press, New York, Chap. 13.
Jennings, B. R. (1974) *Nature* **252**, 33.
Jennings, B. R. (1975) In *Molecular Electro-Optics*, Vol. 1, Ed. C. T. O'Konski, Dekker Co., New York, Chap. 8.
Johnson, M. L., Yphantis, D. A. and Weiss, G. A. (1973) *Biopolymers* **12**, 2477.
Johnson, P. and Perrella, M. (1971a) *Proc. R. Soc.* B **176**, 445.
Johnson, P. and Perrella, M. (1971b) *Proc. R. Soc.* B **176**, 461.
Josephs, R. and Harrington, W. F. (1967) *Proc. Natl Acad. Sci.* (USA) **58**, 1587.
Josephs, R. and Harrington, W. F. (1968) *Biochemistry* **7**, 2834.
Katz, S. and Schachman, H. K. (1955) *Biochim. Biophys. Acta* **18**, 28.

Kauzmann, W. (1959) *Adv. Prot. Chem.* **14**, 1.
Kellett, G. L. (1971) *J. Molec. Biol.* **59**, 461.
Kim, H., Deonier, R. C. and Williams, J. W. (1977) *Chem. Rev.* **77**, 659.
Kinell, P-O (1959) *Arkiv. Kemi* **14**, 327.
King, T. A. and Treadway, M. F. (1976) *J. Chem. Soc. Faraday* II **72**, 1473.
Kirkwood, J. G. (1949) *Rec. Trav. Chim.* **68**, 649.
Kirkwood, J. G. (1954) *J. Polym. Sci.* **12**, 1.
Kirshner, A. G. and Tanford, C. (1964) *Biochemistry* **3**, 291.
Koenig, S. H. (1975) *Biopolymers* **14**, 2421.
Kraemer, W. P. and Diercksen, G. H. F. (1970) *Chem. Phys. Lett.* **5**, 436.
Kratky, O., Leopold, H. and Stabinger, H. (1969) *Z. Angew. Phys.* **27**, 273.
Kuntz, I. D. (1971) *J. Am. Chem. Soc.* **93**, 514.
Kuntz, I. D., Brassfield, T. S., Law, G. D. and Purcell, G. V. (1969) *Science* **163**, 1329.
Kuntz, I. D. and Kauzmann, W. (1974) *Adv. Prot. Chem.* **28**, 239.
Lamm, O. (1929) *Arkiv. Math. Astron. Fysik.* **21B**, 4.
Latimer, W. H. and Rodebush, W. H. (1920) *J. Am. Chem. Soc.* **42**, 1419.
Laurant, T. C. and Killander, J. (1964) *J. Chromatogc* **14**. 317.
Lee, B. and Richards, F M. (1971) *J. Molec. Biol.* **55**, 379.
Lee, J. C. and Timasheff, S. N. (1974) *Biochemistry* **13**, 257.
Liddle, P. F., Jacobs, D. J. and Kellett, G. L. (1977) *Anal. Biochem.* **79**, 276.
Lifshitz, E. M. (1956) *Sov. Phys. JETP* **2**, 73.
Lim, T. K., Baran, G. J. and Bloomfield, V. A. (1977) *Biopolymers* **16**, 1473.
Lloyd, P. (1974) *Optical Methods in Ultracentrifugation, Electrophoresis and Diffusion*. Oxford University Press, Oxford.
Loh, E. (1979) *Biopolymers* **18**, 2569.
Loh, E., Ralston, E. and Schumaker, V. N. (1979) *Biopolymers* **18**, 2549.
Lomant, A. J. and Fairbanks, G. (1976) *J. Molec. Biol.* **104**, 243.
London, F. (1930) *Z. Phys.* **63**, 245.
London, F. (1937) *Trans. Faraday Soc.* **33**, 8.
Long, G. L. and Kaplan, N. O. (1973) *Arch. Biochem. Biophys.* **154**, 696.
Lord, R. C. and Yu, N. T. (1970a) *J. Molec. Biol.* **50**, 509.
Lord, R. C. and Yu, N. T. (1970b) *J. Molec. Biol.* **51**, 203.
McCallum, M. A. and Spragg, S. P. (1972) *Biochem. J.* **128**, 389.
McCammon, J. A., Deutch, J. M. and Felderhof, B. U. (1975) *Biopolymers* **14**, 2613.
McKenzie, H. A. (Ed.) (1971) *Milk Proteins*, Vol. 2, Academic Press, New York and London, p. 358.
McLachlan, A. D. (1964) *Molec. Phys.* **7**, 381.
McLachlan, A. D. (1965) *Disc. Faraday Soc.* **40**, 239.
Magde, D. (1976) *Q. Rev. Biophys.* **9**, 35.
Marshall, A. G. (1978) *Biophysical Chemistry: Principles, Techniques and Applications*, John Wiley & Sons, New York and London.
Mason, M. and Weaver, W. (1924) *Phys. Rev.* **23**, 412.
Mathews, B. W. (1977) In *The Proteins*, 3rd edn, Vol. 3, Eds. H. Neurath, R. L. Hill and G. L. Baeder, Academic Press, New York and London, p. 403.
Mauldin, J. and Fisher, W. R. (1970) *Biochemistry* **9**, 2015.
Miller, D. (1962) *J. Biol. Chem.* **237**, 2135.
Miller, D. (1974) *Biochim. Biophys. Acta* **359**, 172.
Miller, D. B., Summers, M. R. and Niziolek, J. A. (1971) *Nature, New Biology* **230**, 117.
Minton, A. P. (1973) *J. Molec. Biol.* **75**, 559.
Mire, M. (1969) *Biochim. Biophys. Acta* **181**, 35.
Moore, W. J. (1972) *Physical Chemistry*, 5th edn, Longman, London.
Morawetz, H. (1972) *Adv. Prot. Chem.* **26**, 243.
Moser, P., Squire, P. G. and O'Konski, C. T. (1966) *J. Phys. Chem.* **70**, 741.
Munk, P. and Cox, D. J. (1972) *Biochemistry* **11**, 687.

Murthy, A. S. N. and Rao, C. N. R. (1968) *App. Spect. Rev.* **2**, 69.

Myhill, J. (1967) *Biophys. J.* **7**, 903.

Myhill, J. (1968) *J. Theor. Biol.* **23**, 218.

Nagasawa, M. and Takahashi, A. (1972) In *Light Scattering from Polymer Solutions*, Ed. M. B. Huglin, Academic Press, London and New York.

Nakajama, H. and Wada, Y. (1977) *Biopolymers* **16**, 875.

Nemethy, G. and Scheraga, H. A. (1962) *J. Chem. Phys.* **36**, 3382.

Nichol, L. W., Jeffrey, P. D. and Milthorpe, B. K. (1976) *Biophys. Chem.* **4**, 259.

Nichol, L. W. and Winzor, D. J. (1972) *Migration of Interacting Systems*, Clarendon Press, Oxford.

Nozaki, Y. and Tanford, C. (1971) *J. Biol. Chem.* **246**, 2211.

O'Konski, C. T. (1960) *J. Phys. Chem.* **64**, 605.

Olson, T., Fournier, M. J., Langley, K. H. and Ford, N. C. (1976) *J. Molec. Biol.* **102**, 193.

Oncley, J. L. (1941) *Ann. N. Y. Acad. Sci.* **30**, 433.

Oncley, J. L. (1942) *Chem. Rev.* **30**, 433.

Pain, H. J. (1976) *The Physics of Vibrations and Waves*, John Wiley & Sons, Chichester.

Pals, D. T. F. and Hermans, J. J. (1952) *Rec. Trav. Chim.* **71**, 469.

Parsegian, V. A. and Brenner, S. L. (1976) *Nature* **259**, 632.

Pauling, L. (1928) *Proc. Natl. Acad. Sci.* **14**, 349.

Pauling, L. (1960) *The Nature of the Chemical Bond*, Cornell University Press, Ithaca, New York.

Pearce, T. C., Rowe, A. J. and Turnoch, G. (1975) *J. Molec. Biol.* **97**, 193.

Pecht, I. and Lancet, D. (1977). In *Chemical relaxation in Molecular Biology*, Eds. I. Pecht and R. Rigler, Springer-Verlag, Berlin, Heidelberg and New York, p. 306.

Pecora, R. (1968) *J. Chem. Phys.* **48**, 4126.

Perrin, F. (1934) *J. Phys. Rad. Ser. VII* **5**, 497.

Pethig, R. (1979) *Dielectric and Electronic Properties of Biological Materials*, John Wiley & Sons, Chichester.

Perutz, M. F. (1970) *Nature* **228**, 726.

Perutz, M. F. (1975) *Nature* **255**, 256.

Phelps, C. F. (1974) In *Chemistry of Macromolecules*, (Vol. 1 of M.T.P. International Review of Science), Ed. H. Gutfreund, Butterworths, London, p. 325

Phelps, R. A. and Putnam, F. W. (1960) In *The Plasma Proteins*, Ed. F. W. Putnam, Academic Press, New York, Chap. 5.

Phillies G. D. J. (1975) *J. Chem. Phys.* **62**, 3925.

Pimentel, G. C. and McClellan, A. L. (1960) *The Hydrogen Bond*, W. H. Freeman, San Francisco.

Pimentel, G. C. and McClellan, A. L. (1971) *Rev. Phys. Chem.* **22**, 347.

Pitz, E. P., Lee, J. C., Bablouzian, B., Townend, R. and Timasheff, S. N. (1973) *Meth. Enzymol.* **27D**, 209.

Plummer, H. and Jennings, B. R. (1969) *J. Chem. Phys* **50**, 1033.

Porath, J. (1963) *Pure & Appl. Chem.* **6**, 233.

Provencher, (1975) *Biophys. J.* **16**, 27.

Pusey, P. N. (1974) In *Photocorrelation and Light Beating Spectroscopy*, Eds. H. Z. Cummins and E. R. Pike, Plenum Press, New York.

Pusey, P. N., Kappel, D. E., Schaefer, D. W., Camerino-Otero, R. D. and Koenig, S. H. (1974) *Biochemistry* **13**, 952.

Pusey, P. N. and Vaughan, J. M. (1975) *Dielectric and Related Processes*, Vol. 12, Ed. M. Davies, The Chemical Society, London, p. 49.

Raj, T. and Flygare, W. H. (1974) *Biochemistry* **13**, 3336.

Ramachandran, G. N. (1943) *Proc. Indian Acad. Sci.* A **18**, 90.

Raman, C. V. (1959) *Lectures on Physical Optics*, Part 1, Indian Academy of Science, Bangalore, p. 160.

Rao, M. S. N. and Kegeles, G. (1958) *J. Am. Chem. Soc.* **80**, 5724.

Ratner, M. A. and Sabin, J. R. (1973) In *Wave Mechanics. The first* 50 *Years*, Eds. W. C. Price, S. S. Chissick and T. Ravensdal, Butterworth, London, p. 72.
Ree, F. H. and Hoover, W G. (1964) *J. Chem. Phys.* **40**, 939.
Ree, F. H. and Hoover, W. G. (1967) *J. Chem. Phys.* **46**, 4181.
Reinhardt, W. P. and Squire, P. G. (1965) *Biochim. Biophys. Acta* **994**, 566.
Reisler, E. H. and Eisenberg, H. (1969) *Biochemistry* **8**, 4572.
Reisler, E. H. and Eisenberg, H. (1971) *Biochemistry* **10**, 2659.
Reisner, A. H. and Rowe, J. (1971) *Anal. Biochem.* **41**, 1.
Reynolds, J. A. and Tanford, C. (1970) *J. Biol. Chem.* **245**, 5165.
Reynolds, J. A. and Tanford, C. (1976) *Proc. Natl Acad. Sci.* (*USA*) 73, 4467.
Rice, S. A. and Gray, P. (1965) *The Statistical Mechanisms of Simple Liquids* (Vol. 8 of Monographs in Statistical Physics and Thermodynamics), Interscience, New York, p. 112.
Rice, S. A. and Nagasawa, M. (1961) *Polyelectrolyte Solutions*, Academic Press, London and New York.
Richards, E. G., Teller, D. C. and Schachman, H. K. (1968) *Biochemistry* **7**, 1054.
Rigler, R. and Ehrenberg, M. (1976) *Q. Rev. Biophys.* **9**, 1.
Rinde, H. (1929) *Dissertation* University of Uppsala, as quoted in Fujita (1961).
Roark, D. E. (1976) *Biophys. Chem.* **5**, 185.
Roark, D. E. and Yphantis, D. A. (1969) *Ann. N. Y. Acad. Sci.* **164**, 245.
Robinson, R. A. and Stokes, R. H. (1959) *Electrolyte Solutions*, Butterworth, London.
Ross, P. D. and Minton, A. P. (1977) *J. Molec. Biol.* **112**, 437.
Rowe, A. J. (1977) *Biopolymers* **16**, 2595.
Schachman, H. K. (1957) *Meth. Enzymol.* **1**V, 32.
Schachman, H. K. (1959) *Ultracentrifugation in Biochemistry*, Academic Press, New York
Schachman, H. K. and Edelstein, S. J. (1966) *Biochemistry* **5**, 2681.
Schaefer, D. W. and Berne, B. J. (1975) *Biophys. J.* **15**, 785.
Scholte, Th. (1968) *J. Polym. Sci.* A2 **6**, 91.
Schulz, E., Jaenicke, R. J. and Knoche, W. (1976) *Biophys. Chem.* **11**, 253.
Shapiro, A. L., Vinuela, E. and Maizel, J. V. (1967) *Biochem. Biophys. Res. Commum.* **28**, 815.
Smith, G. D., Kirschner, M. W. and Schachman, H. K. (1973) *Biochemistry* **12**, 3801.
Smith, J. and Spragg, S. P. (1979) *Biophys. Chem.* **9**, 215.
Sophianopoulis, A. J. and van Holde, K. E. (1964) *J. Biol. Chem.* **239**, 2516.
Spragg, S. P. (1967) *Anal. Chim. Acta* **38**, 137.
Spragg, S. P., Barnett, W. A., Wilcox, J. K. and Roche, J. J. (1976) *Biophys. Chem.* **5**, 45.
Spragg, S. P., Roche, J. J. and Wilcox, J. K. (1977) *Biochem. J.* **163**, 543.
Squire, P. G. (1964) *Arch. Biochim. Biophys.* **107**, 471.
Squire, P. G., Moser, P. and O'Konski, C. T. (1968) *Biochemistry* **7**, 4261.
Svedberg, T. (1925) *Kolloid-Z. Zsigmondy-Festschrift. Erg. Bd.* zu. **36**, 53.
Svedberg, T. and Fåhraeus, R. (1926) *J. Am. Chem. Soc.* **48**, 430.
Svedberg, T. and Pederson, K. (1940) *The Ultracentrifuge*, Oxford University Press, Oxford.
Tanford, C. (1961) *Physical Chemistry of Macromolecules*, John Wiley & Sons, New York.
Tanford, C. (1973) *Hydrophobic Effect; Formation of Micelles and Biological Membranes*, John Wiley & Sons, New York.
Tanford, C. (1978) *Science* **200**, 1012.
Tanford, C., Buzzell, J., Rands, D. and Swanson, S. (1955) *J. Am. Chem. Soc.* **77**, 6421.
Tanford, C., Kawahara, S. and Lapanje, J. (1967) *J. Am. Chem. Soc.* **89**, 729.
Tang, L. H. and Adams, E. T. (1976) *Fed. Proc. Fed. Amer. Soc. Exp. Biol.* **30**, 1303.
Teller, D. C. (1973) *Meth. Enzymol.* **27**, 372.
Teng, Ta-Lee, Edelstein, C., Barbeau, D. L. and Scann, A. M. (1977) *J. Biol. Chem.* **252**, 8634.
Texter, J. (1978) *Prog. Biophys. Molec. Biol.* **33**, 83.

Thomas, J. O. and Butler, P. J. G. (1977) *J. Molec. Biol.* **116**, 769.
Thomas, J. O. and Edelstein, S. J. (1972) *J. Biol. Chem.* **247**, 7870.
Timasheff, S. N., Dintzis, H. M., Kirkwood, J. G. and Coleman, B. D. (1957) *J. Am. Chem. Soc.* **79**, 782.
Timasheff, S. N. and Inoue, H. (1968) *Biochemistry* **7**, 2501.
Timasheff, S. N. and Townend, R. (1961) *J. Am. Chem. Soc.* **83**, 470.
Timasheff, S. N. and Townend, R. (1970) In *Physical Principles and Techniques of Protein Chemistry*, Vol. 1, Ed. S. J. Leach, Academic Press, London and New York, p. 147.
Tinoco, I. (1955) *J. Am. Chem. Soc.* **77**, 3476.
Tombs, M. and Peacocke, A. (1975) *The Osmotic Pressure of Biological Macromolecules*, Oxford University Press, Oxford.
Trautman, R. and Breese, S. S. (1960) *Anal. Biochem.* **1**, 291.
Trautman, R. and Crampton, C. F. (1959) *J. Am. Chem. Soc.* **81**, 4036.
Trautman, R., Spragg, S. P. and Halsall, H. B. (1969) *Anal. Biochem.* **28**, 396.
Triebel, H. (1972) *Anal. Biochem.* **49**, 198.
van Holde, K. E. (1971) *Physical Biochemistry*, Prentice-Hall, Englewood Cliffs, N.J.
Visser, J., Deonier, R. C., Adams, E. T. and Williams, J. W. (1972) *Biochemistry* **11**, 2634.
Wan, P. J. and Adams, E. T. (1976) *Biophys. Chem.* **5**, 207.
Ward, D. N. and Arnott, M. S. (1965) *Anal. Biochem.* **12**, 296.
Weber, G. (1953) *Disc. Faraday Soc.* **13**, 33.
Weber, K. and Osborn, M. (1969) *J. Biol. Chem.* **244**, 4406.
Wilcox, J. J. (1978) PhD Thesis University of Birmingham.
Williams, J. W. (Ed.) (1963) *Ultracentrifugal Analysis* Academic Press, New York.
Williams, R. C. (1976) *Biophys. Chem.* **5**, 19.
Williams, R. C. and Kim, H. (1976) *Biochemistry* **15**, 2207.
Williams, R. C., Yphantis, D. A. and Craig, L. C. (1972) *Biochemistry* **11**, 70.
Winzor, D. J. and Scheraga, H. A. (1963) *Biochemistry* **2**, 1263.
Winzor, D. J. and Scheraga, H. A. (1964) *J. Phys. Chem.* **68**, 338.
Wodak, S. J. and Janin, J. (1978) *J. Molec. Biol.* **124**, 323.
Wyman, J. (1963) *Adv. Prot. Chem.* **19**, 252.
Yphantis, D. A. (1960) *Ann. N.Y. Acad. Sci.* **88**, 586.
Yphantis, D. A. (1964) *Biochemistry* **3**, 297.
Yphantis, D. A. and Roark, D. E. (1972) *Biochemistry* **11**, 2925.
Zimm, B. H. (1948) *J. Chem. Phys.* **16**, 1099.
Zimm, B. H. and Bragg, J. K. (1959) *J. Chem. Phys.* **31**, 526.
Zipp, A. and Kauzmann, W. (1973) *Biochemistry* **12**, 217.

Index

Activity coefficient
concentration, 25, 55, 61
diffusion, 55
electrolytes, 26
equilibria, 25
hydrophobicity, 10
Albumin
diffusion, 76, 109, 144
radii, 110
relative molecular mass, 76
rotation, 142
sedimentation coefficient, 76
Aqueous solutions, 7
Anisotropy, fluorescence, 145
Association
analysis of boundaries, 162, 180
artefacts, 180
charge effects, 23, 109
classification, 168
in diffusion, 114, 116, 120, 122, 174, 179, 182
dynamic relaxation, 183
errors in coefficients, 155
estimation of parameters, 155
mathematical models, 153, 177
medium energy, 173, 179
qualitative tests, 166
small ligands, 179
stoichiometry, 177, 180
strong interactions, 168
T4, 121
transport, 158, 163, 169, 180, 182
coefficient values, 174
virial coefficients, 23, 154, 157, 162, 177, 181
weak interactions, 181
non-ideality, 154, 158, 162, 182
Autocorrelation, 114, 128, 136
experimental conditions, 138
Averages
in association, 153, 164, 168, 180
relative molecular mass, 49, 55, 77, 154, 155

Base-line, ultracentrifuge, 66, 85
Birefringence, rotary diffusion, 142
Boundary analysis
biological activity, 125, 170
chromatography, 71, 182
diffusion, 124
ultracentrifuge, 84, 162, 180
Brownian motion, 99, 106, 141, 182
light scattering, 126, 128, 138

Charges, shielding, 21, 109
Chemical potential, 10, 13, 14
Chromatography, 69, 182
Coherence area, 131
Concentration
physiological, 43
ultracentrifuge, usable levels, 85, 165, 179
Concentration dependence
albumin $S_{20,w}$, 77, 90
Brownian motion, 99
diffusion, 55, 108, 114, 115, 119, 125, 134, 144, 161
encounter distance, 186
frictional coefficient, 108, 115, 119
myosin, diffusion, 116, 161
rotary diffusion, 122, 143
sedimentation, 77, 79, 110, 119, 158, 162, 164, 169, 174, 186
sedimentation coefficient, 53, 90, 158, 160
theta conditions, 22, 104
theories, 54, 115, 119
tobacco mosaic virus, 90, 174, 180
unshielded polyampholytes, 21
Correlation functions, 130, 134
in diffusion, 131, 134
Correlator, clipped, 137

Cross linking
  bifunctional reagents, 166, 180, 188
  intermolecular, 167, 173

Denaturation, 52, 66, 74, 171
Density, 16, 51, 62, 107, 170
Dichroism, 141
Diffusion (*see also* Rotary diffusion, Translational diffusion)
  associating systems, 114, 116, 120, 122, 161, 179, 182
  biological activity, 125
  concentration dependence, 55, 108, 114, 115, 119, 125, 134, 144, 161, 186
  constant, 49, 78, 106, 133, 140
  correlation functions, 130, 131, 138
  filamentous viruses, 135
  frame of reference, 115, 149
  large particles, 134
  reaction rates, 185
  z-average, 55, 134
Dipole interactions, 32, 146
Dispersion energies, 35, 119
Donnan effects, 17, 117
  osmotic pressure, 14, 17
  ultracentrifuge, 66
Dynamic relaxation, 183, 186
  light scattering, 188
  pressure, 64, 183, 186
  temperature, 183, 186

Electrolytes, 26, 51, 109, 117, 169, 174
Electrophoresis, 41, 73, 138
Electrostatic interactions, 36, 109, 119, 169
Encounter distance, 120, 185
Energy, 2, 139
  absolute versus relative, 24
  dipole, 33, 35
  dispersion, 35
  equilibrium, 24
  molecular interaction, 2, 28, 117, 119, 147, 166, 168, 173
  molecules in solution, 35, 115
  polarizability, 34, 115
  relaxation times, 104, 106
  standard, 24, 25
  weak, 2, 3, 4, 115, 181
Enolase, 152
Entropy, aqueous solutions, 8, 9
Equilibrium
  association, 153, 163, 164, 183, 187
  chemical, 23, 68
  dynamic fluctuations, 28, 187
  energetics, 42
  sedimentation, 56, 177, 180
  time, 57
    analysis, 57, 80, 83
    heterogeneity, 79, 158
  temporal averages, 27, 97, 104
Errors, interaction coefficients, 155

Fick's equation, 124
Flexing, 102, 104, 135, 139
Fluorescence
  autocorrelation, 138
  energy transfer, 146
  life times, 143
Forces
  electrostatic, 29, 54, 119, 173
  gravitational, 29
  hydrodynamics, 54, 119
  hydrogen bond, 31
  ionic, 29
  long-ranging, 2, 37, 119
  molecular, 28, 37, 173
  polyampholytes, 29
  van der Waal's, 28, 32, 35, 38, 54, 119
  weak in biology, 28
Foreignness, molecular, 1
Frictional coefficient, 73, 96, 100
  concentration dependence, 115
  irregular shapes, 101
  molecular mass, 49
  rotational movement, 101
  solvation, 107
  translational diffusion, 55, 106, 115

Gel chromatography
  boundary shapes, 69, 70
  denaturing solvents, 70
  differential boundaries, 71, 182
Gel electrophoresis, 73, 84
Gibb's–Duhem relationship, 13
  partial molar functions, 9
Glutamate dehydrogenase, diffusion, 120

Haemoglobin
  cellular concentrations, 8
  differential gel chromatography, 72
  dimers, 169, 172
  dissociation of dimers, 172
  dissociation of tetramer, 3, 152, 164, 176
  equilibrium centrifugation, 165
  interaction coefficient, 176
  non-ideality, 43, 111

pH, 30, 174
radius of interaction, 116
salt effects, 72, 173
thermally stable, 3
translational diffusion, 111
virial coefficient, 61, 111
Heterodyne light scattering, 135
Heterogeneity
light scattering, 48, 133
limits of detection, 84
moment analysis, 77, 83
quantification, 77
recognizing, 153
translational diffusion, 133
ultracentrifugal boundaries, 55, 78, 162
ultracentrifugal equilibrium, 79, 164
Histones, cross linking, 167
Homodyne light scattering, 128, 135
Hybridization, 171, 172
Hydrodynamic shapes, 20, 98, 100
Hydrogen bond, 31, 32
covalent contribution, 32
in macromolecules, 32
relative energy, 32
temperature dependence, 12
water, 8, 9
Hydrophobic effects, 3, 9, 173
Hydrophobicity
energetics, 9
iceberg, 8
pH, 31
scale, 11
solubility, 12, 13
van der Waal's energies, 36

Ideal solution, Raoult, 15
Insulin, association, 175
Intensity fluctuation spectroscopy, 136
Interactions
molecular, 4, 116, 181
solute–solute, 54, 115, 119, 134, 144, 161, 186
Interaction coefficients
haemoglobin, 115, 176
myosin, 116
systematic errors, 178
Ionic forces, 29, 54, 119, 173

Lactate dehydrogenase
hybridization, 171
solvation, 170
species changes, 152
tetrameric association, 169
β-lactoglobulin, association, 163, 174
Lamm equation, 50, 164
Light absorption, ultracentrifuge, 67, 85, 165
Light scattering
angular dependence, 89, 133
association, 114, 116, 120, 122, 174, 179, 182
autocorrelation, 128
dynamic relaxation, 188
elastic, 46
electrophoresis, 138
graphical analysis, 48, 89
heterogeneity, 48, 80, 120, 134
homodyne, 122
intensity, 45
large particles, 87, 134
non-ideality, 48, 120
quasi-elastic, 126
diffusion, 130
interference equations, 127
large particles, 134
experimental, 138
procedures, 135
temporal autocorrelation, 136
Linear macromolecules, cross links, 67
Lipoproteins, 40
association apo-A1, 160
pressure, 65
Lorentz–Lorenz relationship, 46
Lysozyme, Raman scatter, 140

Macromolecules
fragility, 40
cross links, 67
polymerization, 67
solution, 12, 28, 29, 142
synthesis, 67
Molecular coordinates, 96, 114
Molecular shapes, 97, 114
techniques, 98
Molecules, weighing, 39
Muscle, van der Waal's forces, 37
Myosin
association, 64, 116, 161, 187
translational diffusion, 116

Non-ideality, 1, 8, 61, 89, 109, 117, 176
activity coefficient, 10, 25, 55
biology, 7
heterogeneity, 81, 154
interaction coefficients, 154, 156
Nucleic acids
fractionation, 39
fragility, 40

Nucleic acids, (*contd.*)
  non-ideality, 22, 62, 118, 176
  radius of gyration, 22, 89, 117, 146

Oligomeric complexes
  evolution pressures, 151, 188
  solvent, 30, 171
  species variation, 152, 174
Osmotic pressure, 43
  diffusion, 14, 111, 148
  haemoglobin, 43, 111
  ideal solution, 15
  low concentrations, 43
  neutral molecules, 14
  non-ideality, 44, 111

Paratactic surfaces, 4
Partial specific volumes, 51, 95, 170
pH
  haemoglobin, 30, 178
  inter-particle forces, 36, 109
Phosphoglycerate mutase
  association, 17, 175, 180
  species variations, 152
Photon correlation spectroscopy,
    definition, 135
Polarizability
  dielectric relaxation, 141
  frequency, 34
  light dispersion, 46
  van der Waal's energies, 34
Polyelectrolytes
  activity coefficients, 26
  Donnan effects, 17, 18
    ultracentrifuge, 66
  forces in solution, 29
  partial specific volumes, 51, 117
  virial coefficient, 21, 117
Polysaccharides, 40
Pressure, 14, 43, 64, 183, 186
Proteins
  association, 120, 151, 174
  gel-electrophoresis, 74
  partial specific volumes, 52, 95
  Raman scatter, 139
  virial coefficients, 61, 111, 116, 161
Pyruvate dehydrogenase, concentration
    dependence, 90

Quasi-eleastic light scattering, 126
  autocorrelation, 128
  diffusion, 130
  electrophoresis, 138
  large particles, 134
  procedures, 135, 136
  terminology, 135
  temporal autocorrelation, 136

Radius, interactive, 116, 120, 138
Raman scatter, 139
Raoult's law, ideal, 14, 15
Rayleigh scattering, 46
Refractive index
  dielectric relaxation, 142
  increments, 53
  light scattering, 46, 131
  ultracentrifuge, 66
Relative molecular mass
  absolute methods, 42
  apparent, 20, 61, 177
  averages, 83, 154
  elastic light scatter, 47
  heterogeneity, 75, 82, 154
  interaction coefficients, 154, 174
  gel chromatography, 69
  large particles, 87
  matrix methods, 68
  methods, 41
  osmotic pressure, 16, 43
  relative methods, 41
  sodium dodecyl sulphate, 52
  solvent activity, 42
  ultracentrifuge, 49, 55, 86, 88
  weight average, 48, 154
  z-average, 55, 134, 154
Relaxation
  techniques, 187
  times, 103, 104, 132
    harmonic oscillator, 105
    random coil, 104
    rate constants, 162, 183
    rotary diffusion, 105, 142, 146
Regular solutions, 15
Riboadenylic acid, theta condition, 22
Ribosomes, association, 187
t-Ribonucleic acid
  concentration dependence in diffusion,
    117
  hydrodynamic volume, 118
Rinde equation, 56, 178
Rotary diffusion
  concentration dependence, 143
  dichroism, 144
  dielectric relaxation, 140
  membrane, 146

Sedimentation
  coefficient

albumin, 76
association curves, 159
averages, 77, 79, 155
boundary analysis, 162
concentration dependence, 53, 155, 158
tobacco mosaic virus, 90
corrections, 50
differential boundaries, 161
pyruvate dehydrogenase, 90
constant, 49, 93, 169
equilibrium, 56, 164, 177, 180
Archibald method, 58
experimental analysis, 57, 80, 83
overspeed, 58, 180
temperature, 59
time, 57
non-ideality, 53, 61, 77, 82
velocity (*see also* Ultracentrifuge), 50, 54, 77, 79, 93
association, 158, 162
enzymic activity, 170
pressure, 64
separation, 159
Shapes, hydrodynamic, 20, 98, 101, 117
Sodium dodecyl sulphate
gel electrophoresis, 74
partial specific volume, 52
ultracentrifuge, 66
Solubility
pressure, 12
temperature, 11
Solute interactions, 2, 10, 72, 115, 119
Solvation
estimation, 7, 107
frictional coefficients, 99, 107, 110, 117
hydrophobicity, 10, 72
lactate dehydrogenase, 171
transport, 53, 99, 170
Spheres, concentration dependence, 54, 119
Standard state, 24
Starch, fragility, 40
Structure, theta state, 22, 104
Surface tension, 13
Svedberg equation, 50, 93

T2, frictional coefficient, 102
T4, diffusion, 121
Temperature instabilities, sedimentation, 59
Temporal averaging, 27, 136
Thermodynamics, independence of models, 7
Theta condition, 22, 104
Tobacco mosaic virus
association of protein, 174, 180
arrays in solution, 36, 147
concentration dependence of $S_{20,w}$, 90
rotational relaxation time, 104
Translational diffusion (*see also* Diffusion)
average, 55, 133, 134
concentration dependence, 55, 108, 114, 115, 119, 125, 134, 161
equivalences, 132
frames of reference, 115, 149
frictional coefficient, 100, 106, 115
glutamate dehydrogenase, 120
haemoglobin, 111
inaccuracies, 124
isotopic, 132
mass transfer, 123
myosin, 116
osmosis, 111, 148
T4, 121

Ultracentrifuge (*see also* Sedimentation), 49
association, 164, 174, 180
baselines, 66, 85
boundary analyses, 84, 162
centre of mass, 50
electronic schlieren, 68
equilibrium
equation, 50, 94
heterogeneity, 79, 177
moment analysis, 83
protein loss, 67
velocity equation, 50, 93
velocity mode, 49

van der Waal's
equation, 18
virial expansion, 16
forces, 32, 119
distance, 33
hydrodynamics, 33
large particles, 36
tobacco mosaic virus arrays, 36
Virial coefficients, 1, 16
association, 23, 111, 154, 157, 162, 177, 181
asymmetric particles, 20
calculation, 20, 111, 117
charged molecules, 19, 21, 117
energies, 19
excluded volume, 19, 20, 40, 111, 117

Virial coefficients, (*contd.*)
  gas laws, 16
  haemoglobin, 44, 114
  heterogeneity, 80, 154, 177
  light scattering, 48
  non-ideality, 16, 18
  osmotic pressure, 16, 111
  polarizability, 20, 36
  polyelectrolytes, 18, 21, 117
  proteins, 61, 154, 157, 162, 177, 181
  second, 16, 18, 19, 21, 44, 48, 61, 81, 89, 108, 111, 117, 154, 157, 162, 177, 181
  sedimentation, 53, 61, 77, 82
  translational diffusion, 109, 114, 115, 119, 125, 162
  van der Waal's equation, 18, 36
  Zimm plot, 88
Virus
  electrostatic forces, 30, 36, 147
  filamentous, 135
Volume
  albumin, 110
  apparent partial specific, 51
  electrolytes, 27, 117
  hydrodynamic, t-ribonucleic acid, 118
  macromolecules, 20, 105, 119, 149
  partial, molar, 15, 27, 34, 48
    specific
      amino acids, 95
      centrifugation, 50, 64
      polyelectrolytes, 51, 117
      proteins, 51
  polarizability, 20, 34
  solubility, 12
  solvated molecule, 20, 54, 99, 107, 110, 120
  van der Waal's, 19
  water, 8

Weak associations, 181
Weight average, relative molecular mass, 56, 153, 164, 168, 180

z-average, diffusion, 55, 134
Zimm plot, 88, 133